Sexual Violence on the Jacobean Stage

Sexual Violence on the Jacobean Stage

Karen Bamford

St. Martin's Press
New York

ISBN 0-312-21976-8

Library of Congress Cataloging-in-Publishing Data
Bamford, Karen.
 Sexual violence on the Jacobean stage / Karen Bamford.
  p cm.
 Includes bibliographical references and index.
 ISBN 0-312-21976-8 (cloth)
 1. English drama—17th century—History and criticism. 2. Sex crimes in
literature. 3. Rape victims in literature. 4. Violence in literature. 5. Women
in literature. 6. Rape in
literature. I. Title.
PR678.S4 B36 2000
822'.309355—dc21   99–087928
                                                                   CIP

Design by Letra Libre, Inc.

First edition: May, 2000
10 9 8 7 6 5 4 3 2 1

For R. C.,
Constant Reader

Contents

Acknowledgements   ix

Introduction   1

1. The Legends of the Saints   25

2. Latter-Day Saints   33

3. The Classical Paradigm: Lucrece and Virginia   61

4. "Some Injury in a Matter of Women":
    Variations on the Classical Theme   81

5. Redeeming the Rapist   123

Conclusion   155

Notes   163
Works Cited   211
Index   231

Acknowledgements

P art of chapter 5 was originally published in a slightly different form as "Imogen's Wounded Chastity" in *Essays in Theatre/Études Théâtrales* 12.1 (1993). Part of chapter 6 first appeared, also in a slightly different form, as "Sexual Violence in *The Queen of Corinth*" in *Other Voices, Other Views: Expanding the Canon in English Renaissance Studies*, edited by Helen Ostovich, Mary V. Silcox and Graham Roebuck (Newark: University of Delaware Press, 1999), and appears here with the editors' permission. The cover illustration—from the title page of *Phylaster*, 1620 (STC 1681)—is reproduced by permission of the Huntingdon Library, San Marino, California. I am grateful to Alan Jutzi, Curator of Rare Books at the Huntingdon, for his help in obtaining this illustration. The epigraph to chapter 2 from *The Sadeian Woman* is copyright © Angela Carter 1979, and is reproduced by permission of the Estate of Angela Carter c/o Rogers, Coleridge & White Ltd., 20 Powis Mews, London W11 1JN. Thanks are also due to the following for permission to reprint material: The American Philosophical Society, for the epigraph to the Introduction from *The Letters of John Chamberlain*, edited by Norman McLure; A. & C. Black, for quotations from the New Mermaids *The Revenger's Tragedy*, edited by Brian Gibbons; Cambridge University Press, for material from *The Dramatic Works in the Beaumont and Fletcher Canon*, edited by Fredson Bowers et al., and from *The Dramatic Works of Thomas Dekker*, edited by Fredson Bowers; the Folger Shakespeare Library for quotations from *Hengist, King of Kent*, edited by R. C. Bald; International Thomson Publishing, for quotations from Arden editions of *Cymbeline*, edited by J. M. Nosworthy, *Measure for Measure*, edited by J. W. Lever, and *The Poems*, edited by F. T. Prince; and Scholars' Facsimiles, for the epigraph to chapter 1 from Erasmus's *The Comparation of a Virgin and a Martyr*.

I am grateful to Alexander Leggatt for his exemplary supervision of the doctoral thesis on which this book is based and for his subsequent generous encouragement. I would also like to thank Robert Cupido, Anne Lancashire, Jill Levenson, Barbara Todd and three anonymous readers for St. Martin's Press for their helpful comments; Janet Ritch for her translation of poems on King Roderick and the fall of Spain; Rebecca Bach, Evelyn Gajowski, and Phebe Jensen for sharing their work; and Anne Ward, at the Mount Allison Library, for help in obtaining Interloan materials. I am also grateful to everyone who participated in the seminar on "Reconsidering Rape" at the meeting of the Shakespeare Association of America in San Francisco, April 1999, and in particular to Karen Robertson: Although our subject was somber, the occasion was festive. I would especially like to thank Emily Detmer, who generously sent me her as yet unpublished dissertation, "The Politics of Telling: Women's Consent and Accusations of Rape in English Renaissance Drama" (1997)—essential reading for anyone concerned with the representation of sexual violence and female agency on the early modern stage.

My work has been supported by doctoral fellowships from the Social Sciences and Humanities Research Council of Canada, and grants from the Marjorie Young Bell Research Fund at Mount Allison University. The staff at St. Martin's—Maura Burnett (now at John's Hopkins University Press), Kristi Long, Mara Nelson, Ella Pearce and Amy Reading—smoothed the publication process. Jen Simington's intelligent and sensitive copyediting deserves special mention.

For other kinds of help at diverse stages, I am grateful to Joan Alexander, Virginia Appell, Denise Bamford, Iris Bamford, James Bamford, Jim Bamford, Marie Bamford, Lisa Darrach, Angela Kalinowski, Martha McIntosh, Sarah Novak, Helen Ostovich, Sheila Rabillard, Joy Robertson, Susan Shelby, and Deborah Wills. To Robert Cupido—best of editors, best of friends—I owe personal and professional debts past reckoning.

K. B.
Mount Allison University
Sackville, New Brunswick

Introduction ⬡

Mem. To note how many of the Plays are founded on rapes—how many on
unnatural incestuous passions—how many on mere lunacies.

—Coleridge, marginalia on the *Works* of Beaumont and Fletcher

That morning early there was a joyners wife burnt in Smithfeild for killing her
husband. Yf the case were no otherwise than I can learn yet, she had *summum
jus,* for her husband having brawld, and beaten her, she took up a chesill or
such other instrument and flung it at him, which cut him into the bellie,
whereof he died. Another desperat woman comming from her execution cut
her owne childe's throat, alleging no other reason for yt but that she doubted
she shold not have means to kepe yt. The same day likewise another woman
poisoned her husband about Aldgate, and divers such like fowle facts are committed dayly, which are yll signes of a very depraved age and that judgments
hang over us.

—John Chamberlain, July 6, 1616

Although a study of sexual violence on the Jacobean stage might
plausibly include a variety of aggressive acts, such as the murder of
Desdemona or the torture of Chapman's Tamyra, I have confined
my scope to rapes and attempted rapes. This limitation allows a clear focus
on the significance of chastity—threatened, preserved or lost—for the stage
heroine. I have not, however, attempted a comprehensive analysis of sexual
assault in the drama: The very frequency of the motif would make such a
project unwieldy. Nevertheless, placing well known works by Shakespeare,
Middleton and Heywood beside uncanonized texts, I examine a wide range
of plays in various genres, written over several years, for diverse companies
and audiences. Given the central place of the discourse of chastity in early
modern culture, a study of sexual violence has implications that reach far

beyond the plays I consider; for chastity, though centered on the female body, is—in Eve Kosofsky Sedgwick's apt phrase—a matter "between men."[1]

This project will, perhaps inevitably, appear to some readers—even some feminists—to emphasize unduly the victimization of women. I share their desire to recognize women as agents (I too prefer Webster's Vittoria to Jonson's Celia); and I sympathize with the spirit of resistance in which one might interpret, for instance, the silence of Isabella at the end of *Measure for Measure* as subversive.[2] However, in Jacobean drama issues of female victimization are inseparable from those of female agency. As Emily Detmer observes, since a patriarchal culture requires women to withhold consent to sexual relations from men *other* than their husbands, early modern representations of rape construct women's agency as at once necessary and limited.[3] Moreover, as I argue, in Jacobean drama the victim of rape is, like the witch, invested with an excessive, threatening agency. I use the word "victim" not to insist on female powerlessness, but because its root (*OED* 1) carries a resonance of sacrifice appropriate to the role of the violated heroine.

"Sexual assault" is a modern term, with no Jacobean equivalent. Like "sexual violence," it refers to a range of sexually abusive acts including rape and attempted rape, and I use it—as well as those terms—primarily for that reason. Traditionally, however, the discourse of rape has focused narrowly on vaginal penetration.[4] The jurist Sir Matthew Hale (1609–76) succinctly expressed the perspective of the law on this matter: "To make a rape there must be an actual penetration or *res in re* . . ." (1: 628). This is, as Mackinnon remarks, "a very male point of view on what it means to be sexually violated" (87); and it is one generally endorsed by Jacobean drama, where the vaginal "purity" of the heroine is a matter of life and death. (Thus Shakespeare's Imogen can suffer Iachimo's specular violation and live, but Fletcher's Lucina, who suffers "an actual penetration," dies.)

Historically, and until very recently, rape has been defined in law as a crime of property, in which the rapist was understood to have stolen the value of a female's sexuality from her male proprietor. (Etymologically "ravish" and "rape" derive from the Latin *rapere,* to seize, carry off.) As Deborah Burks observes, the rape statutes were "designed to redress a wrong committed against a woman's male relatives. These men, rather than the woman herself, are considered to be the victims of rape" (765). Thus *The Boke of Justices of Peas* (1506) declares: "[R]ape is where a man ravisheth or taketh a manes wife wydowe or mayde aye[n]st her will & hath to do with her ayn[n]st her will [. . .] " [B4]. This definition turns on the lack of the female's consent ("ayn[n]st her will"), but still conceives of the offence in terms of male property ("a manes wife wydowe or mayde"), even if the male in question is dead.

According to the medieval treatises of Glanvill and Bracton, rape occurred when a man had forcible intercourse with a female illegally—that is, with any female to whom he was not married (J. Carter 37). Bracton's description of the process by which a victim of rape was to appeal the crime presupposes a violent physical attack:

> She must go at once and while the deed is newly done, with the hue and cry, to the neighbouring townships and there show the injury done to her to men of good repute, the blood and her clothing stained with blood, and her torn garments. And in the same way she ought to go to the reeve of the hundred, the king's serjeant, the coroners, and the king's sheriff. (qtd. J. Carter 94)

The law thus required immediate and bloody proof that the victim had not consented to sexual intercourse.

Subsequent medieval legislation, designed to protect the property of wealthy families from undesirable marriages, defined rape and abduction interchangeably (Bashar 30). As Burks argues persuasively, some of this legislation also attempted to "modify the sexual behavior of women," by penalizing those who consented either before or after their rape (766).[5] By the late sixteenth century, however, the two offenses were legally distinct, and the term "rape" had resumed the sexual specificity it carried in Bracton's treatise (Toner 94–95). Sir Matthew Hale, writing in the mid-seventeenth century, describes rape as "the carnal knowledge of any woman above the age of ten years against her will, and of a woman-child under the age of ten years with or against her will" (1: 628). Hale's definition, without immediate reference to a male proprietor, might suggest that the conception of rape was beginning to shift from that of a crime against property to one against the person.[6] However, as Erickson observes, in early modern England women were in many ways still "treated as a form of property" (233), and in the eyes of the law the significance of their violation derived from their status as property or potential property.[7] Here again the law's masculine bias was, as I argue, shared by Jacobean playwrights: The representation of sexual assault on the stage is fundamentally governed by the conception of women as property, and shadowed by the same anxieties about female desire and consent that troubled early English lawmakers (Burks 776).

My starting point is a feminist understanding of sexual assault as a function of culture, rather than of nature; of a socially constructed male dominance, rather than a biologically determined impulse. As the anthropologist Peggy

Reeves Sanday observes trenchantly, "the sex drive does not channel behaviour in the absence of social encoding" ("Rape" 98).[8] Without assigning a single, trans-historical meaning to sexual assault, I have assumed that modern studies, and in particular feminist theories of sexual violence in the twentieth century, provide a critical vantage point for examining its representation in Jacobean drama.[9]

Since the 1970s feminists have analyzed with increasing sophistication the "social encoding" that fosters a rape-prone culture. Pioneering studies by Griffin (1971) and Brownmiller (1975) posit a simple relationship between rape and patriarchy. For Griffin rape is "the symbolic expression of the white male hierarchy" and "the quintessential act of our civilization" (*Rape* 24). For Brownmiller rape is the cornerstone of patriarchy: "From prehistoric times to the present [. . .] rape has played a critical function. It is nothing less than a conscious process of intimidation by which *all men* keep *all women* in a state of fear" (5). In response to this elision of patriarchy and rape by Griffin, Brownmiller and other radical feminists, Roy Porter (1986) poses the question of rape's historical meaning.[10] While acknowledging that rape has recently become "heavily politicized" in North America, Porter argues that it was neither political nor prevalent in traditional society, which "was so securely a man's world that it did not have much need for the 'surplus repression' that endemic rape threats would have conferred" (223). Although Porter's picture of traditional society is contentious,[11] his essay raises important questions about the relationship between patriarchy and rape: "[I]f rape is essentially not sexual release but a crime of violence, can it really lie in the interests of patriarchs or the patriarchal state, to encourage lawlessness?" (234). He concludes that:

> Within Western history [. . .] rape has flourished mainly on the margins [. . .]. Rape has also erupted on the psycho-margins, amongst loners, outsiders, who fail to be encultured into normal patriarchal sex. Above all, rape has marked youthful subcultures, on the criminal margins. [. . .] These young men are not yet absorbed into patriarchy, with its classic roles of husband and father; they lack the permanent erections of mature patriarchy—wealth, property, office, "standing." [. . .] Rapists are thus the waste of patriarchy, but they are its wayward sons not its shock troops; not its life-blood but a diseased excrescence. It should not be forgotten that Western law and morality have abhorred rape [. . .], and that even if its punishment has indeed been sporadic, it has been severe. (235)

Though no apologist for patriarchy, Porter effectively challenges feminists to undertake "the serious business of understanding the ways hierarchical, gendered societies actually operate," and thus "to explain the real resilience of patriarchy" (235).[12]

Porter's challenge is apt. In response, I would emphasize the contradiction that he acknowledges: "Western law and morality have abhorred rape," but its punishment has indeed been "sporadic" at best. At the same time that rape has been condemned with horror, allegations of rape have been treated with scepticism, succinctly expressed by Sir Matthew Hale in an influential dictum: "It is true rape is a most detestable crime, and therefore ought severely and impartially to be punished with death; but it must be remembered, that it is an accusation easily to be made and hard to be proved, and harder to be defended by the party accused, tho never so innocent" (1: 635). Surely more significant than the severity of the punishment for rape in England—where it was a capital offence from 1285 to 1840—is the infrequency with which legal punishment has been exacted.[13]

Although there is no way of determining the actual incidence of rape in Jacobean England, it seems certain that most sexual assaults were not prosecuted in law. In her study of sample records between 1558 and 1700, Bashar found that rape "usually constituted less than 1 percent of all indictments" (34).[14] In addition to the humiliation involved in appealing the crime, the victim would in many cases have been silenced by pressure from, variously, the assailant, his family, her own family and the community.[15] The 1631 case of Margery Evans and Philibert Burghill—discussed in detail by Marcus (293, 296–313) and more briefly by Gossett (313)—provides a clear example of a community's reluctance to prosecute a man accused of rape. According to the evidence presented by Marcus, Burghill and his male servant were almost certainly guilty of raping and robbing the fourteen-year-old Evans. Nevertheless, when she accused them Evans was arrested, physically abused, detained for two days, and soon after imprisoned again at Burghill's suit for almost a month without a charge. The subsequent history of the case reveals the general collusion of the community with the offender. A more public instance of intimidation, supported by some part of the community, occurred in 1615 at Malmesbury in Wiltshire:

> A certain John Vizard, on the eve of his examination before the justices for rape and defamation, showed his contempt for the law and terrified the constable of the town by organizing a parade of armed men with rough music and a mock marriage ceremony, proclaiming that the morrow was his wedding day and bidding company to see him married—"which company [. . .] should be none but cuckolds and cuckold makers."[16]

Of those men actually indicted for rape few were found guilty.[17] The very low rate of conviction may have been due in part to the difficulty of proving rape, in part to the reluctance of (male) juries to take the life of the accused.[18] The figures also suggest, however, both a pervasive skepticism about rape accusations and a general tolerance of the offence.[19]

In spite of the relative silence of court records, I suspect that rape did not flourish, as Porter contends, "mainly on the margins" of early modern English society. In her study of 34 rape prosecutions (1640–1700), Chaytor found that just less than 30 percent of those accused were "gentlemen, yeomen or members of a profession" (403n10). Like Philibert Burghill, these men enjoyed to some extent what Porter calls "the permanent erections of patriarchy—wealth, property, office, 'standing'"; and, like Shakespeare's Angelo, they apparently used the power this gave them to sexually abuse the complainants, none of whom "were of equivalent status or wealth" (Chaytor 403n10). Social theorists suggest that "cultures can and do generate predispositions to behaviour that are, at the same time, defined as deviant or dysfunctional" (Scully and Marolla 307). Though most men in early modern England would no doubt have agreed with Sir Matthew Hale that "rape is a most detestable crime," their culture may well have generated a predisposition to sexual aggression in males. To call rapists "a diseased excrescence" of patriarchy, as Porter does, obscures the possibility that rape—in the seventeenth century and the twentieth—may "represent one end of a quasi-socially sanctioned continuum of male sexual aggression" (Scully and Marolla 306).

My primary concern, however, is with the representation, not the actuality, of sexual assault. Whatever its role in Jacobean society—whether central, or, as Porter argues, marginal—it figures prominently in the drama of the period. Representations of rape are part of its historical meaning; analyzing them is part of the "serious business of understanding the ways in which hierarchical, gendered societies actually operate." While a study of the drama can tell us nothing about the actual incidence of sexual assault—how many were committed, and why—it does reveal the way in which rape was mythologized at a time of social crisis, a time when, whatever the reality of women's power, men certainly did not perceive their own authority as secure. What kind of rape stories did the patriarchal society of Jacobean London entertain itself with?

"Plays [. . .] founded on rapes"

In Jacobean drama sexual assault is—at least on one level—represented as the unruly action of "natural" male desire for female beauty. Here the plays share the sexual discourse of early modern popular culture, in which, as Walker observes, the "language for describing male sexual misbehaviour was that of ordinary, male, heterosexual activity. Rape was depicted as an extreme expression of men's 'lustfull Desires' and 'pleasures'" (5). Indeed, Jacobean

drama amply demonstrates what Brownmiller calls the myth of "the beautiful victim" (370–80). All of the assaulted heroines are, like Jonson's Celia,
eminently desirable—young, beautiful, chaste, and well-born—and almost
all of their assailants are, like Volpone, ostensibly driven by lust.[20] Because
her beauty "provokes" the attack, the heroine implicitly shares responsibility
for the aggression. (Thus the exemplary Celia begs for mutilation as an alternative to rape, imploring Volpone to "punish that unhappy crime of nature, / Which you miscall my beauty—flay my face, / Or poison it with
ointments for seducing / Your blood to this rebellion" [3.7.250–53].) In
general, too, like Volpone, the villains are clearly villainous: Monsters of depravity, vicious tyrants—whether in the public realm, like Middleton's Duke
of Florence, or in the domestic realm, like Tourneur's D'Amville—they are
reassuringly different from ordinary men.[21]

As the word "ravish"—with its connotations of overwhelming pleasure
(*OED* 3c)—suggests, the common belief that women enjoy being raped operates throughout these plays as a largely unspoken but effective assumption.[22] Similarly, sexual assault appears as a coercive invitation to sensual
pleasure that a good woman adamantly refuses. It thus acts as an index, or
test, of her chastity: The chaste woman is *most* chaste when she is saying "no"
to the sensual pleasure with which her assailant tempts her. Accordingly, the
struggle between victim and attacker is constructed as a contest between
virtue and vice. Castabella refuses the sexual pleasure D'Amville explicitly
offers her (4.3.99–124); Celia says no to Volpone's "sensual baits" (3.7.209);
and Milton's Lady rejects the cup of pleasure Comus holds out to her.

In tension with the mystifying discourse of virtue and vice, however, another pattern emerges—a pattern reflecting a patriarchal structure in which
women, defined as male property, act as tokens of exchange between men.[23]
Rape, on this level, is something men do to *other men's* property. As *The Boke
of Justices of Peas* puts it: "[R]ape is where a man ravisheth or taketh *a manes
wife wydowe or mayde* aye[n]st her will & hath to do with her ayn[n]st her
will [. . .]" (emphasis added). Like cuckoldry, rape thus involves a triangular relationship between assailant, victim and her male proprietor(s). In illegally possessing a female, the rapist dishonors and dominates another man,
or men, as well as the victim. His sexual aggression is thus an aspect of what
Breitenberg calls his "anxious masculinity," a subjectivity dependent upon
"conquest and acquisition," and "coincident" with desire (125); and one that
necessarily involves competition with other men. Assault thus functions as a
sign of the woman's extraordinary value in an economy of competitive male
desire: It marks her as a prize. Volpone desires the unknown Celia *because*
she is Corvino's jealously guarded possession; her chastity itself is eroticized,
and his attempt on that chastity is part of his larger campaign to rob and hu-
miliate her husband. Here, as in the model of mimetic desire first theorized

by René Girard, the female object is less important than the antagonistic relationship between the male rivals.[24]

Something of this triangular dynamic of desire, rivalry and aggression is suggested by the woodcut from the 1620 Quarto of Beaumont and Fletcher's *Philaster*, reproduced on the cover of this book. In the foreground the victorious "Cuntrie Gentellman" jubilantly flourishes his phallic sword, in possession of the stage and potentially in possession of the reclining "Princes" as well. After defeating Philaster, here pictured skulking in the foliage, the anonymous gentleman—or "Country Fellow"[25]—cries to Arethusa, "I pray thee, wench, come and kiss me now" (4.5.107–8), implying her status as sexual prize. Like a Spenserian heroine, Arethusa, who has been wounded by Philaster in an erotic stabbing, is briefly at the mercy of her rescuer.[26] Since the immediate arrival of a courtly search party ensures that the Princess is in no danger, however, the loutish demand of the country "boor" (103) is a source of comedy, not suspense.[27] Nevertheless, the woodcut offers a convenient image of a recurrent triangle, in which the female functions primarily as prize and object of desire, and male-male combat is foregrounded.

Carolyn Williams observes that in Shakespeare's work "the violated, silenced female body" functions "as a middle term in a transaction between men" (94). Even when the heroine successfully resists assault, in Jacobean drama sexual violence is always represented within a web of male-male relationships; or, in Rebecca Ann Bach's useful phrase, within "the homosocial imaginary" that structured the culture of early modern England.[28] Celia's threatened chastity is "a middle term" in transactions between Volpone and—severally—Mosca, Corvino, Bonario, Corbaccio, Voltore and indeed the Commandatori, the "fathers" who represent the political hierarchy of Venice. Since, as Bruce Smith contends, the "all-male power structure of sixteenth-and seventeenth-century society fostered male bonds above all other emotional ties" (56), it is not surprising that in dramatic fictions sexual aggression against women *matters* primarily as it affects these privileged male bonds, both enabling and disrupting homosocial solidarity. Typically the sexual aggression polarizes the male community. Initially an occasion for male bonding between the assailant(s) and allies—such as Chiron, Demetrius and Aaron in *Titus Andronicus*—it strengthens an opposing homosocial alliance among the supporters of the injured husband/father—such as Marcus, Titus and his sons—and sparks a political conflict.[29] Volpone's attempt on Celia generates comedy in part because, contrary to the logic of male honor, it unites rather than divides Volpone and Corvino, and disrupts rather than strengthens the bond between Bonario and Corbaccio. Missing from most of these plays is any sense of a female community for the assaulted heroines: Like Lavinia and Celia, they are defined almost exclusively through their relations with men.[30] Within the "homosocial imaginary" the raped woman presents a problem to her community: In part because she signifies a potentially explosive grievance between men, in part because she herself is dangerous. In the classical myth of Lucretia—from a patriarchal perspective, a "best-case scenario"— the suicide of the violated woman empowers her husband and his allies, provoking a rebellion that purges the state: This fiction of male aggression and female sacrifice appealed powerfully to the early modern imagination, and it appears in various guises on the Jacobean stage. However, in more than one Jacobean analogue the political conflict provoked by the rape is represented as socially destructive, and such ironic variations of the Lucretia myth reveal an anxiety about the rape victim at its heart. In the contrasting classical legend of Philomela the raped woman joins with her sister to exact a terrible revenge, including parricide and cannibalism.[31] If the Lucretia story idealizes the self-destructive rape victim, the Philomela story, expressing a powerful "dread over the violence possible to women's revenge" (Lamb 216), demonizes the vengeful rape victim.

The work of feminist critics like Joplin, Lamb and Jane Newman on Lucretia and Philomela helps to explain some of the anxieties surrounding the representation of sexual violence in Jacobean drama, where the victim of rape is both idealized (like Lucretia) and demonized (like Philomela).[32] Although the self-destructive Lucretia is the favorite classical model for the sexually violated heroine, the specter of the angry Philomela haunts these plays. For, though few of the rape victims take revenge into their own hands, all are nevertheless represented as dangerous. In the anthropological terms proposed by Mary Douglas, the violated woman has been forced into an ambiguous social position, and is thus invested by her community with "uncontrolled, unconscious, dangerous, disapproved powers—such as witchcraft and evil eye" (Douglas 99). Sexually "impure" herself, she is seen as a potential source of pollution not just to her bloodline, but to the community at large. "As the sign and currency of exchange," Joplin argues, "the invaded woman's body bears the full burden of ritual pollution. [. . .] If marriage uses the woman's body as good money and unequivocal speech, rape transforms her into a counterfeit coin, a contradictory word that threatens the whole system" (41–42). This sense of pollution is figured, in Jacobean drama, through language of disease and infection. The victim's anger compounds the danger she represents. As Douglas observes of witches, it is "the existence of an angry person in an interstitial position which is dangerous" (102). Abused and alienated, the raped woman is "an angry person in an interstitial position" and a source of danger to her community. Like

Ovid's Medusa, she is punished with a monstrosity that signifies the threat she embodies.[33]

The tensions between the mystification of sexual assault as a moral struggle, in which the assailant *tempts* his victim to illicit sensual pleasure, and the underlying construction of sexual assault as theft, in which the assailant *steals*, or tries to steal, property, generate contradictions in the pervasive discourse of chastity. According to this discourse a woman's chastity is at once an immaterial virtue and a material possession; it is simultaneously a moral code of vigilant self-discipline and a commodity that may be stolen from her in spite of her best efforts; it is both a mode of conduct (the chaste woman *behaves* chastely) and a physical state (the chaste woman is *vaginally* chaste). The heroine who loses her chastity in the second sense (physically) is perceived as unchaste in the first sense; that is, deficient in virtue, no matter how she has behaved.[34] She loses her chastity if, under any circumstances, *with or without her consent*, a man who is not her husband has "carnal knowledge" of her. Without exception, she understands herself to be infectious and to have suffered a radical loss of identity (she typically calls herself a "whore").

In a discussion of Massinger's plays, Otten poses what she calls a "troubling question": "Why does rape in Massinger lead so surely to death?" (146). She finds the answer in the symbolic value of "the rape-death connection," which she argues "represents the opening and closing of ancient wounds, and is sacrificial [. . .] [I]t stands as emblem of [the victim's] powerlessness, the destruction of her illusions about her protectedness" (146). Otten's question *is* troubling, and it invites an answer that goes beyond the symbolism of rape and death in Massinger alone to address the larger discourse of chastity that dominates the representation of women on the early modern stage. In other words, Otten's question can (and should) be asked not just of Massinger's work, but of all early modern drama. For unless the stage victim of rape is able to marry her assailant, and thus regain her chastity, she dies after her violation. No other means of reintegration into the patriarchal social structure is allowed her: She must either live chaste (by marrying the rapist) or die unchaste.

Understandably disturbed by the tragicomic solution of marriage to the rapist, Suzanne Gossett connects this "decadent Jacobean exploration" (327) of the rape plot with the decadent morality of the Jacobean court.[35] She concludes:

> We cannot be satisfied with either outcome, the woman who must die once she is raped or the woman "happily" married to her attacker. Paradoxically, the

plays which assume that rape victims must die may imply a concern and respect for women in general which is missing from the plays which do not au-
tomatically condemn the heroine to death. (326–27)

Although Gossett identifies an interesting shift in the way rape is treated on
the Jacobean stage—the plots do become increasingly diverse—I believe the
change she notes reflects a quest for dramatic novelty rather than the "increasingly conspicuous" vice at court (305).[36] More significantly, however,
Gossett's argument obscures the extent to which earlier plays—as she describes them, "based on classical models and influenced by English law and
traditional morality"—are themselves ambivalent about rape. These classically derived plays do not, I believe, imply "that women have a personal integrity that cannot survive violation" (324); nor do they "imply a concern
and respect for women in general." As much as the "marriage" plays, the
"traditional" plays represent an attempt to accommodate rape. More significant than any variation in plot is the discourse of chastity common to all
the plays, early and late—a discourse enshrined in the classical legend of Lucretia as in Christian hagiography, codified in English law and endorsed by
"traditional morality."

History, Gender and the Drama

As Coleridge and others have observed, many Jacobean plays are "founded
on rapes." The obvious explanation for this phenomenon—that such violence, then as now, was commercially successful—is unsatisfying. It simply
provokes the question: *Why* was this motif so popular with audiences and
playwrights at this historical moment? Even if some plots (the war story, for
example) seem to have a perennial appeal in western culture, it does not
mean that appeal is always constant (war stories have been especially popular in America since the Vietnam war). How does the representation of sexual assault on the Jacobean stage relate to the meanings of gender in
Jacobean society? As McLuskie observes,

> in the case of theatre, the direct link between cultural forms and social forms
> is broken by the particular characteristics of the theatre as a mode of cul-
> tural production. [. . .] [T]he material of the theatre [. . .] was part of a
> stock which was recycled and reproduced as much with regard to its saleable
> theatrical potential as in response to the direct pressures of particular cul-
> tural needs. The cultural meaning of "women" on the Elizabethan stage can-
> not, as a result, be inferred by direct recourse to social or ideological
> pressure since it was constantly mediated by the form of the drama and the
> demands of the theatrical institutions which it sustained. (*Renaissance
> Dramatists* 123–24)

Any attempt to determine "the cultural meaning" of the sexual violence that pervades Jacobean drama must acknowledge the texts as commercial products, and the particular dramatic forms the playwrights inherited. We can, however, infer that a diverse group of playwrights writing for diverse audiences in Jacobean London recognized "the saleable theatrical potential" of sexual assault, and we can ask what kind of "social or ideological pressure" created the theatrical market these playwrights were catering to. If, as various critics have suggested, the prominence of women in early modern drama does not reflect a "Renaissance" for actual women, but is rather a register of contemporary anxiety—anxiety which, in Catherine Stimpson's phrase, "attached itself to gender and stories about gender" (K. Newman xii)—we can understand the prominence of sexual violence in the drama as, in part, the theater's commercial response to an appetite sharpened by such anxiety. As I argue, historical evidence offers considerable support for this position.[37]

Economically, women were, on the whole, losing ground throughout the sixteenth and seventeenth centuries. Though historians have disagreed with minor aspects of Alice Clark's analysis, they have largely supported her contention that an emergent capitalism marginalized women and devalued their work.[38] In some areas, such as the silk and brewing trades, women lost an earlier ascendancy (Alice Clark 138–43, 221–29; Cahn 53–57), while in medicine they were overtly excluded (H. Smith, "Gynecology" 108–09; Cahn 58–60). In spite of the presence of a strong female ruler, the role of Elizabethan women was circumscribed more narrowly in the home. Far from encouraging women to move into public life, Elizabeth I reinforced and exploited patriarchal stereotypes (C. Levin, "Advice" 183; Strong 16; Benson 234–35).

The intellectual upheaval of the Renaissance failed to effect a significant change in the scholarly treatment of women (Maclean 1). The medieval scholastic synthesis, in which sex difference was conceived in terms of the Aristotelian dualities—active/passive, form/matter, perfection/imperfection—retained its powerful, conservative hold on the interrelated disciplines of theology, medicine, ethics and law (Maclean 82). Maclean finds, moreover, that underlying this scholasticism were:

> Pythagorean dualities, which link, without explanation, woman with imperfection, left, dark, evil and so on. These emerge most obviously in medicine, but are implied in theology and ethics also. Although they are nowhere explicitly defended, they may nonetheless be the most accurate indicator in anthropological terms of the status of woman in Renaissance society and culture. (87–88)

Analyses of medical texts support Maclean's judgement (H. Smith, "Gynecology" 103–105; Crawford, "Woman's View" 66–67). Legally women remained severely disadvantaged (Stone, *Family* 195; Hogrefe xx-xxiii). Erickson concludes that, in spite of disjunctures between a restrictive theory and a practice that allowed actual women "considerably more power over property than has previously been allowed, both the legal system and individual men still kept women firmly subordinate. Women's dependence on their men's good will increased over the [early modern] period . . ." (17). The author of *The Lawes Resolutions of Womens Rights* (1632) sums up their position succinctly. After recounting the biblical story of the fall, he declares:

> See here the reason [. . .] that Women have no voyse in Parliament, They make no Lawes, they consent to none, they abrogate none. All of them are understood either married or to bee married and their desires or [sic] subject to their husband, I know no remedy though some women can shift it well enough. The common Law here shaketh hand with Divinitie [. . .]. (6)

Reformed "Divinitie" differed little from Catholic with respect to women. Recent studies suggest that the reformers were at least as patriarchal, in theory and in practice, as their predecessors;[39] and although Protestantism may have promoted education for men, it did not advance that of women (J. Morgan 176–77; Charlton 205–07). Perhaps most crucially, medieval views of sexuality persisted despite the elevation of marriage (Blench 274; Brundage 575; Stone, *Family* 498–501). Indeed, the reformers' opposition to the enforced celibacy of medieval religious life stemmed from their conviction of the radical corruption of human nature, rather than a positive reevaluation of sexuality. Following Augustine, they stressed the strength of carnal passions; following St. Paul, they insisted that "It is better to marry than to burn" (1 Cor. 7. 9). Sexual activity in marriage was thus primarily, in the words of the Elizabethan *Book of Common Prayer,* "a remedy against sin, and to avoid fornication" (290), rather than a positive good. Luther's commentary on *Genesis* suggests the extent to which even marital sex remained a source of anxiety:

> [I]t is a great favour that God has preserved woman for us—against our will and wish, as it were—both for procreation and also as a medicine against the sin of fornication. In Paradise woman would have been a help for a duty only. But now she is also, and for the greater part at that, an antidote and a medicine; we can hardly speak of her without a feeling of shame, and surely we cannot make use of her without shame. The reason is sin. In Paradise that union would have taken place without any bashfulness, as an activity created and blessed by God. [. . .] Now, alas, it is so hideous and frightful a pleasure that physicians compare it with epilepsy or falling sickness. [. . .] We are in the state of sin and of death; therefore we also undergo this punishment, that

we cannot make use of woman without the horrible passion of lust and, so to speak, without epilepsy. (Works 1: 118–19)[40]

Reformed theologians thus agreed with Catholics in condemning strong sexual passion within marriage (I. Morgan 149–50; Stone, *Family* 200). As Calvin expressed it, "though honourable wedlock veils the turpitude of incontinence, it does not follow that it ought forwith to become a stimulus to it. Wherefore, let spouses consider that all things are not lawful for them" (Bk. 2, c. 8; 1: 350).[41]

The Protestant marriage remains the subject of fierce disagreement. Stone's thesis of the "reinforcement of patriarchy"—anticipated to some extent by Hill and Walzer—is supported by the work of several scholars.[42] Some historians, however, like MacFarlane and Houlbrooke, doubt the impact of Protestant marriage doctrine on actual behavior.[43] Others, reacting against the "new history" of Ariès and Stone, have written in defense of the patriarchal marriage and family.[44]

Challenging the celebration of "the Puritan art of love" in the Hallers' pioneering studies, Halkett's analysis of Puritan marriage doctrine emphasizes its conservative nature (14–20). "Puritan writers," he declares, "are not distinguished by any special concept of the relationship between husband and wife" (24).[45] Davies confirms the essentially traditional nature of Protestant teaching on marriage, but also notes an extraordinary emphasis on obedience in the handbooks. She finds a "Puritan obsession with achieving male dominance" and observes that the "overwhelming preoccupation of the seventeenth-century writers was with the relationship which subordinated the wife to the husband" (63).[46] Similarly, Karen Newman notes that Protestant texts like Whately's *A Bride-Bush* "represent, by bringing into discourse the husband's power over his wife's body, an obsessive, even self-conscious *display* of masculine agency" (10). Although, as Lena Orlin argues, "female power was *in practice* imperfectly contained by the deployment of an ideology relentlessly committed to the maintenance and articulation of male authority" (103), prescriptive writers expended considerable ingenuity trying to reconcile female domestic roles with notional male supremacy. Certainly, the proliferation of conduct books in the sixteenth century gave old gender roles a new authority.[47]

Whatever the balance of power in actual marriages, it seems certain that the issue of sovereignty was the source of widespread concern. Like Davies, Walzer notes an emphasis on authority. In the Puritan literature of family life, he comments, "neither nature nor love played much of a part. Their concern was almost entirely with the 'government' of the household; they wrote prolix chapters with such titles as 'How women ought to be governed,' and 'How children owe obedience and honor to their parents'" (188). According to William Whately, the popular Puritan divine, the whole duty of a husband lay in the "keeping of his authority, and the using of it":

> Nature hath framed the lineaments of his body to superiority, & set the print of government in his face, which is more sterne, lesse delicate then the womans. He must not suffer this order of nature to be inverted. The Lord in his Word cals him the head; hee must not stand lower than the shoulders; if he doe, that is a deformed family. It is a sin to come lower than God hath set one. It is not humility, but basenes, to be ruled by her whom he should rule. (18–19)

The whole duty of the wife was "to acknowledge her inferiority" and "to carry her selfe as inferiour" (36). Whately and others gave elaborate advice to husbands and wives on how to fulfill these duties. Halkett finds that "the qualities of reverence, obedience, modesty, retirement, silence, patience, and compliance characterize ideal wifely behaviour for most [. . .] matrimonial writers" of the period.[48]

Anglicans too exalted the power of the husband. In the absolutist political theory of early seventeenth century, the parent-child relationship was seen as standing for that of monarch and subject, master and servant, and husband and wife.[49] The Anglican preachers, Walzer writes, "stretched fatherhood to an overbearing absolutism, but did not cease to urge the necessity of mutual love between subject and king" (186). King James's marital advice to his son in his *Basilikon Doron* reveals the same tension:

> Treat her as your owne flesh, command her as her Lord, cherish her as your helper, rule her as your pupill, and please her in all things reasonable; but teach her not to be curious in things that bellong to her not: Ye are the head, shee is your body; It is your office to command, and hers to obey [. . .]. (*Political Works* 36)

Anxiety about male authority is evident in other areas of Tudor and Stuart society. Underdown's analysis of church court records reveals "an intense preoccupation" with women who were "a visible threat to the patriarchal system," especially between 1560 and 1640: "Women scolding and brawling with their neighbours, single women refusing to enter service, wives dominating or even beating their husbands: All seem to surface more frequently than in the periods before and afterwards" ("Taming" 119).[50]

A disruptive woman might find herself presented to the court for being "a busy woman of her tongue," "a common scold and a disturber of the whole parish" or even "a maker of rhymes, thereby to raise slanders"; or for "libellous and lascivious ballads" or "inventing nick-names." The woman found guilty might be ducked on the cucking-stool, imprisoned or led through the streets by a metal bridle (K. Thomas, *Religion* 631; Boose 185–90). Unruly wives and/or their displaced husbands might also be subject to the unofficial discipline of the community in the form of the charivari, or skimmington ride.[51] Davis (124–51) and Ingram ("Ridings" 93–99) both stress the festive nature of the charivari. Davis even suggests that it may have been subversive of the patriarchal hierarchy. Perhaps. But its primary characteristic seems to have been punitive, its method violent. Thus, as Ingram reports, in Wiltshire in 1618, Agnes Mills was dragged from her house by a group of three or four hundred men, "thrown into a wet hole, trampled, beaten and covered with mud and filth" ("Ridings" 82). The aim was clearly, at least on one level, social control. A participant in a 1604 riding claimed they intended that "not only the woman which had offended might be shamed for her misdemeanor toward her husband [in beating him] but other women also by her shame might be admonished [not] to offend in like sort" ("Ridings" 93).

As Underdown points out, the "epidemic of scolding" coincided with the height of witchcraft prosecutions ("Taming" 120). The two crimes are closely related: Both usually began in cursing or railing—indeed, as Keith Thomas notes, "a reputation for successful cursing could easily lead to a formal charge of witchcraft" (*Religion* 610). They are the crimes of the otherwise impotent. In both cases, the accused were overwhelmingly female and poor (*Religion* 620–32).[52] Dolan argues that:

> By attributing power to such [marginal] women, popular culture collaborated with the law in criminalizing poverty. Like fictions of petty treason, representations of . . . witchcraft locate the threat to domestic and social order in the least powerful and privileged, in those most likely to be victims rather than the perpetrators of violence, vilification, and exploitation. (15)

Whatever complex socioeconomic factors underlay the persecution of scolds and witches, the related phenomena suggest a widespread misogyny.[53]

Continental writings on witchcraft, like *Malleus maleficiarum* (1486), were, as Keith Thomas acknowledges, strongly "anti-feminist," dilating upon "the theme of diabolic copulation and the lore of incubi and succubi" (*Religion* 679). According to Couliano the literature of witchcraft borders on the "pornographic": "[T]he inhibitions of an entire era of repression are poured into it. All possible and impossible perversions are ascribed to witches and their fiendish partners" (214).[54] This continental, sexual, idea of witchcraft became current in England at the end of the sixteenth century, disseminated through a series of popular pamphlets and learned treatises, including King James's *Daemonologie* (1597) (K. Thomas, *Religion* 523–24).

Unruly women were, by common agreement, unchaste women (Boose 195–96). As Whately observed of the scolding wife: "This impudencie, this unwomanhood tracks the way to the harlots house, and gives all wise men to know, that such have, or would, or soone will cast off the care of honesty, as of loyaltie" (38). Not surprisingly then, the "epidemic of scolding" and the witchcraft trials coincided with a growing emphasis on female chastity (Pearse 52).

To some extent this insistence on chastity may be seen as part of a larger concern with sexual discipline. Historians have observed a heightened emphasis on sexual morality from the middle of the sixteenth century (Ingram, *Church Courts* 125; Sharpe, *Defamation* 24–25). From 1543, when bills "for true keeping of matrimony," "for the incontinency of women" and "for women lawfully proved of adultery to lose their dower, goods, lands and all other possessions" came before the House of Lords, there were repeated attempts to make adultery a criminal offence (K. Thomas, "Puritans" 273). In 1547 the first volume of the *Homilies* brought Becon's "Sermon Against Adultery" to parishes all over England. The latter half of the sixteenth century also saw the elevation of lechery to "a prominent place among the Seven Deadly Sins" (Pearse 59).[55] Economic considerations buttressed spiritual: A prosperous middle class was laying up treasure on earth, and Puritan moralists emphasized adultery's threat to property as well as the soul (Greaves 230–31). In a discussion of clandestine and irregular marriages in the early modern period, Cressy observes that "the age of Queen Elizabeth and the early Stuarts stands out as a period of exceptional cultural discipline. Legal, social, moral, and religious pressures brought all but the most marginal or the most reckless into line and into church" (*Birth* 316).

One aspect of the "exceptional cultural discipline" described by Cressy may have been the "new interest and urgency" surrounding the question of bodily self-control in early modern European culture (Paster 25). Drawing on the work of Elias, Foucault and Bakhtin, Paster argues that Galenic humoral theory was "instrumental in the production and maintenance of gender and class difference" during this period (7). As she puts it, "the crucial problematic was whether women as a group could be counted on to manage their behaviours in response to historically emergent demands of bodily self-rule" (25). The gendered discourse of Renaissance medical texts suggests that in general women could *not* "be counted on to manage their behaviours" appropriately: Such texts inscribed the female body as dangerously uncontrollable—open, leaky, incontinent.[56]

Although some Puritan moralists challenged the double standard, in practice it was widely accepted (K. Thomas, "Double Standard" 212; Cahn 144; Gowing, "Language" 28). Chastity was the responsibility of women, and a large body of literature—sermons, pamphlets, conduct books, plays and poems—served to remind them of it (Pearse 49–133). Thus Edmund Tilney, in *The Flower of Friendship* (1568), states that a wife must "above all imbrace chastitie. For the happinesse of matrimonie, doth consist in a chaste matrone [. . .]." (D4r). Thomas Bentley, in his *Monument of Matrons* (1582), an encyclopedic devotional book specifically for women, declares: "There is nothing that becometh a maid better than soberness, silence, shamefastness, and chastitie, both of bodie and mind. For these things being once lost, she is no more a maid, but a strumpet in the sight of God" (qtd. Hull 142). Bentley includes model prayers for maidens and matrons, for the preservation of their chastity, and "A lamentation of anie woman, virgin, wife, or widowe, for hir virginite or chastitie, lost by fornication or adulterie" (qtd. Pearse 54).[57] Homiletic drama reinforced the lesson. A common plot in domestic tragedy involved the horrible results of a wife's infidelity (Adams 6–7). In the first of the dumb shows in *A Warning for Fair Women,* the figures of Lust and Chastity vie for the heroine: *"Lust imbraceth her, she thrusteth Chastity from her, Chastity wringes her hands, and departs"* (811–13). Stricken with remorse after the murder of her husband, she laments:

> A womans sinne, a wives inconstancie,
> Oh God that I was borne to be so vile
> So monstrous and prodigious for my lust.
> Fie on this pride of mine, this pamperd flesh,
> I will revenge me on these tising eies,
> And teare them out for being amourous. (1556–61)

As Dolan observes, such "stories of women who plot against their husbands articulate and shape fears of the dangers lurking within the home, of women's voracious and ranging sexual appetites and capacities for violence, and of the instability of masculine privilege and power" (58).

Again, church court records substantiate literary evidence. According to Gowing's study of slander litigation,

> insults of women played on a culpability for illicit sex that was unique to them. The personal, verbal, social and institutional sanctions against "whores" and "bawds" had no counterpart for men. Men were less likely to be presented for illicit sex. Men's adultery was never an accepted ground for marital separation, as women's was. And the word "whore" had no male equivalent. Instead, men fought cases over insults like "whoremonger," "pander," or "cuckold," concerning not their own sexuality but that of women for whom they were in some sense responsible. ("Language" 29–30)

Thus court records reveal a society in which sexual reputation was paramount for women, and not only among the upper classes. The concern was shared by women in the middle and, to a large extent, the lower orders as well (Sharpe, *Defamation* 17). In the explosion of defamation cases after 1560, women—in contrast to men—overwhelmingly complained of sexual slander (Sharpe, *Defamation* 15; Amussen 208; Gowing, "Gender" 2). Gowing concludes that women "were accused in more sexual detail and with more ramifications than men. They were defamed, principally by other women, through a view of female reputation defined in relation to men" ("Gender" 19). Given the legal practice of purgation—by which a defendant brought witnesses who testified to her good reputation—slander could have serious material consequences. Ingram notes the comment of one Yorkshire woman when she heard another defamed: "[T]hey might as well take her life as her good name from her" (*Church Courts* 165).

Gowing's study of "the language of insult" in defamation cases points to a connection between sexual misconduct and a perceived threat to the patriarchal household: "As whoredom damaged the body, so it endangered the corpus of the household. Most defamations directed their complaints at a particular sphere: the effects of fornication, adultery, and bawdry upon household order" ("Gender" 13). According to the common charges, a whore alienated a man's money and affection from his wife, and—if she was married—made her own husband a cuckold. She thus undermined male control of two households:

> The adulterous husband loses control of his money, the cuckolded husband loses control of his female property, and both are consumed outside the household. Extended further, the disturbance of order is visualized as destroying the household itself. One way of identifying a whore or a bawd was apparently her broken windows; some women had their windows broken for them. But it is cuckolds who are most threatening to the framework of the home. The horns imagined upon them are a constant danger. Cuckolds put their heads out of windows and cannot get them in again; they break doors; they threaten to damage the fabric of the neighbourhood. ("Gender" 14)

The anxieties about the loss of male control noted by Gowing also emerge in Orlin's study of household or "oeconomic discourse" in post-Reformation England. Orlin argues that the abdicating or "renouncing" husband, who failed to maintain moral order in the household, was a prototype of domestic evil in the Jacobean years (195).

Anxieties about gender roles also emerged throughout the period in a formal controversy over female nature. Elaborate, usually conventional works attacking or defending women proved a highly marketable commodity.[58] As Woodbridge points out, however, the "feminism" of the defence is illusory:

Attacks and defenses disagree, or pretend to disagree, about whether there are more bad than good women in the world; but their definitions of female goodness and badness are identical. The formal defense ultimately accomplished little more than to proffer a stereotype of the good woman to complement the attack's stereotype of the bad. (*Women* 133)[59]

Hester, Sara and Grissell thus opposed Jezebel, Delilah and Medea. Above all, however, defenders of women celebrated Susanna, Penelope and Lucrece. These three, more than any others, exemplified the cardinal female virtue of chastity (Pearse 71).[60]

"If the formal controversy had any purpose at all beyond literary delight," Woodbridge observes, "then the purpose of attacks and defenses was likely the same—to enforce a certain mode of behavior" (*Women* 134).[61] The debate may thus be seen both as a register of social anxiety about the behavior of women, and an instrument of behavior modification. Beginning in the 1540s, the controversy continued intermittently until the early years of James's reign, when Louis B. Wright notes "a gradual intensifying of acrimony" (481), followed by a wave of "especially violent" satire between 1613 and 1625 (484n36). The reasons for this escalation of bitterness are obscure.[62] However, it may be part of a larger cultural pattern. The controversy about women, the widespread anxiety about chastity, the witchcraft prosecutions, the epidemic of scolding, all point to what Underdown has described as "a period of strained gender relations" in the decades around 1600 ("Taming" 136).

It seems unlikely that this crisis was provoked by a real increase in women's independence.[63] More probably it was symptomatic of the general crisis historians have perceived in the period. Victor Harris finds the century between the Reformation and the Civil War characterized by a universal fear of disorder. He traces a general concern with the corruption of nature, "first growing around the middle of the sixteenth century, rising to an extensive and continuous excitement from the 1570's into the 1630's, and subsiding sharply from then on. The peak of this interest comes during the years immediately preceding and following [. . .] 1616" (87). The fear of chaos Harris charts throughout the period is directly related to a moral anxiety: "[T]he complaint is regularly about immorality or irreligion, and the solution is regularly repentance. More and more, from the 1580's through the first half of the seventeenth century, [. . .] the explanation of man's miseries is directed to the improvement of his morals" (109). Harris's research is supported by Walzer's contention that the spread of Puritanism was a response to "an acute fear of disorder and 'wickedness'—a fear [. . .] attendant upon the transformation of the old political and social order" (303). Similarly, Stone has argued that the "authoritarian family and the authoritarian nation state were the solutions to an intolerable sense of anxiety, and a deep yearning for order" (*Family* 217).[64]

The most crucial years seem to be roughly the half-century between 1580 and 1630. It is in these years, Heal observes, "that men seem to be most sharply aware of the threat to traditional values, and most eager to counter that threat with prescriptive advice" (68). It is probably not coincidental that this period of intense anxiety corresponds roughly to the rise in witchcraft and scolding prosecutions; nor that the years in which Harris sees the peak of this fear, the years around 1616, are also those in which the formal controversy about women reaches its bitter height. They are also those in which, according to Stone, the "crisis of the aristocracy" reaches its climax (*Crisis* 198).

Linking witchcraft accusations to the "many uncertainties, ambiguities, tensions and conflicts" generated by the restructuring of society in early modern Europe, Hester argues that the persecution of witches served "as a means of recreating the male status quo in the emerging social order" (289). In a complementary argument, Robin Briggs notes that witchcraft trials "drew communities together to purge themselves of the evil within, and used the idea of pollution to reinforce threatened boundaries" (65). Similarly Stuart Clark, exploring the significance of the order-disorder opposition that preoccupied writers during the period, observes that descriptions of chaos served to reassert the existence of order, that demonology in effect validated Christianity: "Thus James's own attempt in 1590–1 to write into the confessions of the North Berwick witches a special antipathy between demonic magic and godly magistracy had been a way of authenticating his own, as yet rather tentative initiatives as ruler of Scotland" (117). Such observations provide a useful insight into the crisis in gender relations around 1600—the witchcraft prosecutions, the scolding, the charivaris, the formal controversy—and into the representation of women in the drama. As Woodbridge observes, "a group, a culture can have collective needs, collective anxieties, to which group magical thinking can be a response" (*Scythe* 15). As chaos apparently threatened to come again, a patriarchal society, responding to its collective anxieties with "group magical thinking," demonized the unruly woman, as witch, scold or whore.[65] Given the principle of binary opposition that, as Clark points out, "was a distinctive aspect of a prevailing mentality" in this era (105), the demonization of the disorderly woman involved the sanctification of her opposite: The chaste, silent and obedient heroine of the conduct books. From this dynamic emerged the pairs of women who so frequently appear on stage: Tamora/Lavinia, Celia/Lady Wouldbe, Vittoria/Isabella, predator/victim.

We may plausibly relate the prevalence of sexual violence in the Jacobean drama—the large number of plays, as Coleridge put it, "founded on rapes" or attempted rapes—to this "magical thinking" about order and disorder, purity and danger: "[A]kind of magical thinking," as Woodbridge explains, "that lies below the surface of consciousness—the sense that rape (even sex itself) is a pollution, that an anxious and threatened society can magically protect itself by defending (in political symbolism, in literature) the orifices of the body" (*Scythe* 71). Like her binary opposites—the witch, the scold, the whore—the chaste victim of sexual aggression provided a scapegoat for an anxious patriarchal society. The heroine who resists assault, who begs for death before dishonor, who accepts the guilt for the crime, who kills herself afterwards—such an (exceptional) woman legitimizes the patriarchal ideal of chaste womanhood, and thus the patriarchal order. According to the logic of scapegoating, the sexually predatory villain (himself a scapegoat for the larger community) transfers guilt to the heroine, who absorbs and atones for it.[66] A reassuringly pure victim, she is nevertheless punished for the disruptive desire she provokes and for the sex she shares with the disorderly woman—the witch, the scold, the whore—who threatens the patriarchal household.[67]

Obviously, to say this is not to imply a conscious and unified ideological project on the part of Jacobean playwrights. On the contrary. As McLuskie points out, social pressure, such as the crisis in gender relations posited above, will appear only indirectly in the drama, mediated by commercial interests and inherited dramatic forms. I assume a diverse group of dramatists, largely professional or quasi-professional, from varied backgrounds, with varied skills and interests, united perhaps only by their involvement in the London theater, and for the most part concerned primarily with making a living in an intensely competitive market.[68]

I assume too that audience response is gendered—that is, shaped by the spectator's experience as a gendered subject; and that, given the economic and social constraints on women, men made up the majority of the playgoing public in Jacobean England.[69] If, as Richard Levin argues, women "were regarded by the playwrights and acting companies as a constituency whose interests and feelings should be considered" ("Women" 165), in these fictions of sexual violence they were nevertheless offered patriarchal fantasies. When the twice-martyred Lady is carried off for reburial at the end of *The Second Maiden's Tragedy*, the hero commends her as an example to her sex: "I would those ladies that fill honor's rooms / Might all be borne so honest to their tombs" (5.2.211–12). While female spectators of any class were no doubt able to resist such direct ideological pressure, most probably enjoyed imaginative identification with the desired, aristocratic and heroic object of the Tyrant's lust.[70] Male spectators, who must have made up the majority of the audience, probably enjoyed identification with Govianus, the heroic revenger, and perhaps—on another level—with the desiring and transgressive Tyrant.

I trace the contradictory logic of chastity through three groups of plays, connected by the themes of sanctity and sacrifice. In the first are those in which the heroine's resistance to sexual assault, like Celia's, defines her as saintly: Shakespeare's *Pericles*, *The Virgin Martyr* by Massinger and Dekker, Henry Shirley's *The Martyred Soldier,* and the anonymous *The Two Noble Ladies.* These plays illuminate an important stereotype of female virtue and present a paradigm that is implicitly at work in all the plays examined. Like Milton's Lady, the heroines of these plays successfully refuse the cup of sensual pleasure, and their triumphant chastity redeems the men around them—variously persecutors, fathers and suitors. In the second and largest group are plays that either dramatize the Roman legends of Lucretia and Virginia—Heywood's *The Rape of Lucrece* and Webster's *Appius and Virginia*—or work a variation on their pattern of female sacrifice and male aggression, in which the heroine's suffering redeems her community from its bondage to the tyrant-rapist: *The Revenger's Tragedy,* Shakespeare's *Cymbeline,* the anonymous *The Second Maiden's Tragedy,* Fletcher's *Valentinian,* Rowley's *All's Lost by Lust,* and Middleton's *Hengist, King of Kent.* Fletcher's *Bonduca,* in which the ravished females refuse to sacrifice themselves, and the male hero refuses the task of revenge, forms an ironic coda to this group. In the last group are plays in which the rapist, rather than the community, is redeemed: Shakespeare's *Measure for Measure, The Queen of Corinth,* by Fletcher, Massinger and Field, and *The Spanish Gypsy,* attributed to Middleton and Rowley. Here the assault represents a young man's fall into a state of sin from which he is rescued by the providential craft of a benevolent older man and the forgiveness of his victim.

Read individually, these plays might seem merely peculiar, perverse or sensational in their representation of sexual violence. (They include variously necrophilia and scopophilia, torture and mutilation, parricide and suicide, a bed trick, two rape tricks and several threatened gang rapes.) Read collectively, however, they form a mutually illuminating and congruent group: Each rearranges inherited motifs of sexual aggression—male lust, valor and revenge; female purity, sacrifice and pollution—into patterns that map the significance of chastity for the Jacobean stage heroine. They reveal the fundamental role of male property rights in that contradictory and powerful discourse; and they reveal too, the persistent scapegoating of the assaulted woman, who is both idealized and demonized on the Jacobean stage. Finally, I argue, a study of these plays shows the various kinds of cultural work they performed, managing patriarchal anxieties, naturalizing sexual assault and in diverse ways rationalizing it as redemptive.

Chapter 1 ❁

The Legends of the Saints

Virginity [. . .] is not praiseworthy because it is found in martyrs, but because it makes martyrs.

—St. Ambrose, *De Virginibus*

A true virgyn doth differre very lyttell from a martyr. A martir suffreth the executioner to mangle his fleshe: a virgin dayly dothe with good wyll mortifie her fleshe, she beinge in maner a turmentour of her selfe. [. . .] What men have suffred the tourmentes of martyrdom more mervailously and strongly than virgine Martyrs, Agnes, Cecilia, Agatha, and other theyr felowes innumerable? And therefore whan a virgin is delyvred to the executioner, she dothe not begynne her martyrdome, but makethe an ende of that that she beganne longe before.

—Erasmus, *The Comparation of a Virgin and a Martyr*

Although, as I argue, stories of sexual violence had a particular appeal for the Jacobean audience, the sanctification of the victim of assault is ancient. It is, indeed, central to Christian hagiography: In the lives of the virgin martyrs, the Church enshrined the sexually threatened woman as supremely heroic. So important are the saints' lives to Jacobean drama, both for the model of the female heroic and the values that inform that heroism, that an analysis of sexual violence on stage must begin with a brief account of the hagiographic tradition.

Celibacy was the ideal for both sexes in the early Christian Church,[1] but it had special significance for a woman. On one level virginity allowed her a sexual ambiguity. Not yet fully female, because sexually uninitiated, she enjoyed a measure of maleness (Hastrup 58–59; Warner, *Alone* 72–73). She thus to some extent escaped the guilt, as well as the chief penalties—pains of childbirth and domination by a husband—acquired by her sex at the Fall. By practicing a rigid asceticism she could even achieve the accolade of virility.[2] Moreover, as the cult of the Blessed Virgin demonstrates, the virginal female body was a potent symbol of purity and integrity. As Peter Brown observes, the female virgin:

> was the one human being who could convincingly be spoken of as having remained as she had first been created. Her physical integrity came to carry an exceptionally high charge of meaning. To late antique males, the female body was the most alien of all. It was as antithetical to them as the desert was to the settled land. When consecrated by its virgin state, it could appear like an untouched desert in itself: it was the furthest reach of human flesh turned into something peculiarly precious by the coming of Christ upon it. (271)

Precious, but still dangerous. For the Church Fathers the female body, virginal or otherwise, was primarily a source of sexual temptation, and as Tertullian angrily declared, even the chastest of virgins ought to go veiled.[3] Whatever power or dignity women won in the early Church through sexual continence merely mitigated a prevalent and deeply rooted misogyny.[4]

Both the symbolic power of women and the misogynistic theology that determined that power are reflected in the legends of the female saints. Under the pressure of a growing asceticism, the stories of the early women martyrs changed from bare statements of testimony and execution to elaborate tales of heroic chastity.[5] Thus, even though many of the martyrs were matrons, the type of female saint to emerge in the legends is a virgin.[6] Her beauty provokes the carnal passion of a noble or powerful man, who usually offers her marriage, wealth and rank. When she resists his suit, she is prosecuted for her Christianity, tortured and finally executed.[7] Though lengthy tortures are reported for male as well as female saints, there is often a sexual dimension to the women's torments. As Warner remarks, "the particular focus on women's torn and broken flesh reveals the psychological obsession of the religion with sexual sin, and the tortures that pile up one upon another with pornographic repetitiousness underline the identification of the female with the perils of sexual contact" (*Alone* 71). According to De Voragine's popular *The Golden Legend,* Juliana and Barbara are both beaten naked (3: 45, 6: 202–4) and, like Eufemia, hanged or dragged by the hair (3: 46, 5: 146, 6: 202). Agatha is rolled naked on burning brands (3: 37), and her breasts, like Barbara's, are drawn and cut off (3: 35 and 6: 203).[8]

When death finally does come to the martyrs, it is usually by the phallic sword, suggesting that they suffer a symbolic rape. This aspect of the martyrdom is frequently emphasized by diction, as in Aldhelm's description of Lucy's death: "But with a rigid sword he violated her beautiful womb: / Purple blood straightway flowed from the flesh."[9] Frequently "bride of Christ" imagery heightens the eroticism of the killing: The moment of the virgin's death is simultaneously the moment of her ecstatic union with Christ. Prudentius describes the passion of Agnes in terms that render this erotic drive explicit:

> When Agnes saw the grim figure standing there with his naked sword her gladness increased and she said: "I rejoice that there comes a man like this, a savage, cruel, wild man-at-arms, rather than a listless, soft womanish youth bathed in perfume, coming to destroy me with the death of my honour. This lover, this one at last, I confess it, pleases me. I shall meet his eager steps halfway and not put off his hot desires. I shall welcome the whole length of his blade into my bosom, drawing the sword-blow to the depths of my breast; and so as Christ's bride I shall o'erleap all the darkness of the sky and rise higher than the ether. (2: 343)

Here, so eager is the bride for her heavenly spouse, that she longs for the death thrust. The symbolic rape of martyrdom is no longer rape: It is ecstatic coition.[10]

Several of the virgins are threatened with literal rape or rapes. Eufemia's judge follows her into prison, and, in Caxton's words, "would have taken her by force for to have accomplished his foul lust, but she defended her forcibly, and the virtue divine made the hands of the judge to be lame" (De Voragine 5: 145). Later, when various tortures have proved ineffectual, he sentences her to be raped to death ("send to her all the young men that [. . . are] jolly, for to enforce and to make her do their will till she [. . .] fail and die" [De Voragine 5: 146]). Eufemia, however, remains inviolate. Likewise a judge sentences Lucy to death by rape in a brothel ("labour her so much till she be dead" [De Voragine 2: 134]), but the Holy Ghost prevents it. Although Daria reaches a brothel, a lion defends her (De Voragine 6: 610). Agnes and Agatha also emerge from brothels miraculously unscathed (De Voragine 2: 247–48 and 3: 33), as does Irene (*Acts of the Christian Martyrs* 291).

For the women in all of these stories Christianity means primarily chastity: They are dedicated to virginity, consecrated to a heavenly spouse. In some legends there is no formal prosecution on religious grounds, and the primary significance of chastity emerges starkly, unobscured by the apparatus of the judicial contest. Three basic elements remain: The suitor's demand, the virgin's refusal and her murder. As a seventeenth-century English martyrology tells the story of Winifred, the girl is alone in her house when she is visited by a prince who "impudentlie urgeth her to let him have his pleasure of her."[11] Pretending she is going to change her clothes, Winifred escapes through the back door: "The impious Prince hearing that she was so slipt away, runneth presentlie after her, and overtaketh the innocent lambe, and he renewing his former filthie suite, but she denying him, affirming that she was ioyned unto Christ, wherefore she could not, neither would ever couple herself with man, the furious youth raging at her answer, with his sword cuttes of her head [. . .]" (*Lives of Women Saints* 90). The erotic implications of the sword wound emerge clearly in a Welsh version of the tale in which the murderer is described as "dishonouring her with a sword" (Henken 142).

This kind of narrative reaches a bizarre climax in the medieval legend of St. Ursula and the Eleven Thousand Virgins. There are conflicting accounts of how Ursula and her company reached Cologne, but most agree that they were slaughtered there by the Huns.[12] The story in *The Lives of Women Saints* emphasizes the moral purity of the women:

> These HUNNES then beholding the excellent beautie and comlinesse of these women, as they themselves were moste prone to leacherous lust, so did they incite these virgins to the like; wooing them moste egerlie to have their pleasures of them. But holie URSULA, no lesse glorious for the claritie of vertue and virginitie, than for hir nobilitie of birth, instructed all her companie with so sound admonitions of pietie and Christianitie, that they all chose constantlie to suffer death, rather than with detriment of their faith and chastitie, to yield unto the Barbarians fleshlie desire. Thereat the HUNNES, that could not stay in that place [l]ong moved with a great rage, in barbarous cruell manner kill[ed] the whole companie. (38)

Eleven thousand virgins, solicited by a host of Huns, deliberate between the pleasures of the flesh and the dictates of virtue, choose the latter and are slain by the disappointed pagans: The surreal proportions of the story expose the sexual mythology embodied in the lives of the virgin martyrs. What do these legends say about sexual assault? They tell us that it is the natural response of unredeemed masculinity to feminine "beautie and comlinesse," and thus that the female body is the instrumental cause of the assault. They tell us too that if a woman is good enough, the assault will be ineffectual: Frustrated by the virtue of the women, the Huns simply kill them; they do not rape them. Rape is thus implicitly cast as one term in an either/or choice presented to the women (either agree to sexual intercourse or die)—as if it were not more likely to be part of a both/and package (first rape, then death) that they could not refuse. The reality of rape is thus silently denied: If a woman can "freely" choose death instead of sex, then the assault is simply seduction (or a most eager wooing). If she chooses life (and sex), she does so to the "detriment of [. . .] faith and chastitie."

In most cases, as in this version of Ursula's death, the implications of the woman's responsibility for sexual aggression remain hidden. Sometimes, as in Hrotsvitha's tenth-century play *Callimachus,* they become explicit. Threatened with assault, the virgin Drusiana prays:

> O Lord Jesus, what use is my vow of chastity? My beauty has all the same made this man love me. Pity my fears, O Lord. Pity the grief which has seized me. I know not what to do. If I tell anyone what has happened, there will be disorder in the city on my account; if I keep silence, only Thy grace can protect me from falling into the net spread for me. O Christ, take me to Thyself. Let me die swiftly. Save me from being the ruin of a soul! (Plays 55)

Christ hears her prayer and takes her. Drusiana heroically chooses death—in effect, suicide—instead of seeking justice (which would be disruptive) or enduring rape (which would be the destruction of her chastity and her assailant's soul). She acknowledges her responsibility for her plight ("My beauty has all the same made this man love me"), and assumes the guilt and consequent punishment.[13] Simultaneously innocent and guilty, she dies for the sins of another and of her sex.

Officially the Catholic Church has understood virginity to be a spiritual state, depending on "the absence of all voluntary and complete venereal pleasure." Thus "integrity of the flesh is no more than an accidental element," and its involuntary loss through rape leaves virginity intact (Camelot 703).[14] This doctrine, however, is implicitly denied by the hagiographical tradition. From its earliest days, the Church has celebrated women, like Ursula or Drusiana, who choose disfigurement or death to preserve the "accidental element" of physical integrity.[15] In spite of Augustine's judicious disapproval, it encouraged women to embrace death and a mystical bridegroom.[16]

Only rarely in hagiographical accounts is the real possibility of rape explicitly addressed. Ambrose tells the story of a virgin (nameless) who has to choose between denying Christ and being cast into a brothel. Anxious for her chastity, she deliberates and decides that "it is better to have a virgin mind than a virgin body."[17] However, her virginity is providentially preserved in the brothel through a soldier's self-sacrifice. Ambrose triumphantly proclaims the moral of the story: "The virgin of Christ can be exposed to shame, but not contaminated. Wherever the virgin of God is, there is the temple of God. Neither can brothels defame chastity; rather, chastity banishes the infamy of the place."[18] As a symbol of spiritual integrity, this fiction of physical inviolability serves an obvious doctrinal purpose. However, it also subtly reinforces a masculine reluctance to believe in the reality of rape. The corollary to the proposition that God preserves the integrity of the chaste is that the woman who is raped was not truly chaste. Again, the rape is not really rape.

Occasionally the Fathers did acknowledge the reality of rape. Basil of Ancyra had words of comfort for women who might face assault:

> In the time of the persecutions virgins who had been pursued because they remained faithful to the Husband and were given to impious men managed to preserve their bodies from being defiled; for He for whom they had suffered paralyzed and struck with impotence the violence those impious men tried to use on their bodies; and He preserved their very bodies, I tell you, completely free of corruption by an astonishing miracle, or else if they had been taken by violence, as their soul took no part in the fleshly pleasures, they seemed to mock their dead bodies and they gave their soul, which had refused to give in to the sensual pleasure of the one who had outraged them, to their true Husband, pure and more shining in its fidelity and virginity. (qtd. Rousselle 192. Emphasis added.)

Basil's words make clear the logic implicit in the lives of so many female saints. To suffer rape actually *enhances* a woman's "fidelity and virginity" in the eyes of God. (Such enhancement, however, is apparently contingent on physical death; Basil does not address the problems faced by the survivor of rape because survival was clearly not a preferred option.) In the legend of Lucy the "enhancement" principle is again articulated. The girl's angry fiancé declares, "I shall do bring thee to the bordel, where thou shalt lose thy chastity, and then the Holy Ghost shall depart from thee." Lucy defies him: "The body may take no corruption but if the heart and will give thereto assenting [. . .]. And therefore, if thou make my body to be defouled without mine assent, and against my will, my chastity shall increase double to the merit of the crown of glory" (De Voragine 2: 134). This may have been some consolation to victims of rape, but from a masculine perspective it was a dangerous message: It might soften female resistance. The saints' lives, as we have seen, manage to have it both ways: They insist on the physical inviolability of the women while allowing them to suffer assault, erotic torture and symbolic rape. Even though Lucy's physical integrity is supernaturally preserved, the principle she enunciates certainly applies to her and her fellow virgin martyrs. Their bodies *are* "defouled," though not technically violated, by the sexual assaults they endure, and their chastity—their value in the eyes of God and man—increases "double to the merit of the crown of glory."

For the Christian woman, then, the ultimate form of heroism was to endure a sexual assault—and the graver the threat, the greater the heroism—before dying in defence of her chastity.[19] Throughout the Middle Ages girls and women were encouraged to emulate the martyrs as well as the Blessed

Virgin herself. The author of the thirteenth century homily, *Hali Meiden-had*, exhorts his female readers:

> Think of St. Katherine, St. Margaret, St. Agnes, St. Juliana, St. Lucy, St. Cecilia, and of the other holy maidens in heaven; how they not only re-fused kings sons and earls with all worldly wealth and earthly joys, but en-dured strong pain rather than accept them and a sorrowful death at last. Think how well they are off now, and how they revel now in Gods arms as queens of heaven! (63–64)

In the sixteenth century the humanist Vives gave the sexual morality of the saints' lives renewed authority. His influential treatise *The Instruction of a Christian Woman*—a text that circulated widely in Protestant England[20]—celebrates the power of virginity:

> How much, then, ought that to be set by, that hath ofttimes defended women against great captains, tyrants, and great hosts of men? We have read of women that have been taken and let go again of the most unruly soldiers only for the reverence of the name of virginity [. . .]. O cursed maid and not wor-thy to live, the which willingly spoileth herself of so precious a thing which men of war that are accustomed to all mischief yet dread to take away. Also lovers, which be blind in the heat of love, yet they stay and take advisement. For there is none so outrageous a lover if he think thee be a virgin, but he will always[s] open his eyes and [. . .] take counsel to change his mind. (Klein 104–5)

Vives makes the implications of this "reverence" for virginity clear: Since "no man will take [a maid's virginity] from her against her will, nor touch it, ex-cept she be willing herself," the woman who does lose her virginity under any circumstances is an "evil keeper" of her chastity and has only herself to blame (Klein 107). Again, rape is not really rape.

In its rejection of enforced celibacy and in its hostility to what Foxe calls the "idle fantasies and forged miracles" (2: 64) of Roman Catholic hagiography, the Protestant Reformation challenged traditional conceptions of sanctity and disrupted the cult of the saints in England.[21] Married chastity rather than "nunnish virginity" became the ideal state for women. Nevertheless, in his *Acts and Monuments* Foxe commends the heroic chastity of the early fe-male martyrs and includes several tales of their exemplary resistance to sex-ual solicitation and assault.[22] His fiercely Protestant work thus enshrined the stories of many early saints in a widely available and authoritative text.

Moreover, in spite of the missionary zeal of Elizabeth's government, considerable regional and local variations in religious belief remained throughout England in the sixteenth century (Haigh 37–53). A Puritan document of 1584 estimated that "[t]hree parts at least of the people" were "wedded to their old superstition still" (qtd. in K. Thomas, *Religion* 84). Saints' day observance was repressed with difficulty and in some places continued into the next century (Cressy, *Bonfires* 4–33; Hill, *Society* 142–44). Local veneration of parish saints persisted (K. Thomas, *Religion* 81), and despite repeated attempts to suppress it, the cult of St. Winifred flourished at Holywell through the seventeenth century.[23] Images painted on walls, cloths, or in stained glass preserved for generations the iconography of the saints (T. Watt 134–38, 173–74, 202). Significantly, the "great and growing attraction" (Duffy 173) of the legendary virgin martyrs in late medieval England ensured that many of their images survived (171–73). Traditional religious drama continued, as Paul Whitfield White observes, "well into Elizabeth's reign. Some of these plays continued to espouse Catholic teaching on [. . .] the veneration of the saints, and the cult of the Virgin, and were performed under parish, municipal, and academic auspices that remained sympathetic to pre-Reformation religion and sensibilities" (134–35).[24]

Thus, though the Protestant hierarchy campaigned vigorously against such "superstition," traditional hagiography survived—certainly as legend, if not as the object of widespread religious faith.[25] Driven underground, traditional hagiography emerged covertly in popular romances, both narrative and dramatic, like the first part of Deloney's *The Gentle Craft* (1597) and *Sir Clyomon and Sir Clamydes* (c. 1570).[26] At a more courtly level it appeared, suitably transformed, in that most Protestant of poems, *The Faerie Queene,* and in the revised *Arcadia,* where the topos of the virgin martyr receives one of its most sustained treatments (Hannay).

The hagiographical tradition also emerged on the Jacobean stage, where it profoundly affected the representation of female heroism. The virgin martyr sets the implicit standard for the heroine threatened with sexual violence. If, like Jonson's Celia and Marston's Sophonisba, she adheres to the example of the saints, she shares in their glory. If, like Middleton's Bianca and Beatrice-Joanna, she deviates from their model, valuing her life more and her chastity less, her deviation is a sign of moral and spiritual corruption; she is, in Vives's words, "an evil keeper" of her chastity and she is punished for it by vituperation as a whore. In the Jacobean drama, as in the lives of the saints, the *truly* chaste woman is inviolable; her body may be "defouled" by symbolic rape or erotic torture but not sexually violated. Instead her chastity increases with her suffering "double to the merit of the crown of glory."

Chapter 2 ✤

Latter-Day Saints

> All the mythic versions of women, from the myth of the redeeming purity
> of the virgin to that of the healing, reconciling mother, are consolatory non-
> sense; and consolatory nonsense seems to me a fair definition of myth,
> anyway.
>
> —Angela Carter, *The Sadeian Woman*

In general the influence of the hagiographic tradition on the representation of sexual violence in Jacobean drama is pervasive and implicit: Thus
Celia and Sophonisba respond to sexual assault in terms that confirm
their generic sanctity but connect them to no saint in particular. However,
in a small group of plays the hagiographic debt is direct and substantial.

The earliest of these is Shakespeare's *Pericles* (c. 1608).[1] Based on the ancient legend of Apollonius of Tyre, it is self-consciously archaic, recalling the
saint's plays of the Middle Ages.[2] Nevertheless its hagiographic debt is
covert: Marina is a Christian saint in disguise. Between 1618 and 1624,
however, while King James pursued a Spanish bride for Prince Charles, English Catholics enjoyed a marked respite.[3] Henry Shirley's *The Martyred Soldier* (1619), Massinger and Dekker's *The Virgin Martyr* (1620), and the
anonymous *The Two Noble Ladies and the Converted Conjuror* (c. 1623) responded to the shifting political and religious climate by treating overtly hagiographic material.[4]

All these Jacobean "saint's plays" dramatize what Angela Carter calls the
"consolatory nonsense" of a virgin's redemptive purity: Fathers, suitors,
would-be ravishers are born again, converted, made new. Theatrically exciting assaults on the heroines' chastity provoke or provide dramatic crises.[5]
In spite of their similarities, however, each of the plays treats the topos of

sexual assault differently, and the differences are instructive. Shakespeare strengthens the heroine he found in his primary sources, endowing Marina with a consolatory power that emerges clearly when one contrasts her with the heroines of the later plays. In *The Martyred Soldier* Shirley introduces both pathos and sensational episodes of sexual violence into his primary source narrative: Attempted rape becomes a public contest—enacted through the saintly female body—between an anti-Christian tyrant and the Christian God. In *The Virgin Martyr* Massinger and Dekker outdo Shirley in staging sexual assault as a theatrical event. Exploiting the erotic potential latent in their sources, they represent their heroine's suffering with pornographic emphasis. Finally, in *The Two Noble Ladies,* the playwright radically softens the legendary St. Justina and makes her idealized weakness crucial to the conversion of her would-be ravisher.

Pericles

The Apollonius legend, as Shakespeare received it, is about an incestuous sexual assault and its consequences. In all known source versions of the story— the fifth-century Latin *Historia Apollonii Regis Tyrii,* Gower's *Confessio Amantis,* the medieval *Gesta Romanorum* and Laurence Twine's *The Patterne of Painefull Adventures* (1576)—Antiochus rapes his daughter.[6] Twine's account substantially preserves that of the *Historia* and the *Gesta,* in which Antiochus succumbs to the madness of desire: "[T]he furious rage of lust pricking him forward thereunto, he violently forced her, though seely maiden she withstood him long to her power, and threwe away all regard of his owne honestie, and unlosed the knot of her virginitie" (426). Afterwards, in her grief, the daughter decides on suicide ("I can have no remedie now but death onely" [427]), although she is dissuaded by her nurse. Gower too blames Antiochus unequivocally. He concludes, "The wilde fader thus devoureth / His owne flesh, which none socoureth, / And that was cause of mochel care" (317–19).

Traditionally, then, Antiochus's daughter is a sympathetic figure, wholly a victim. Her suicidal response to the attack is one encouraged by a patriarchal society. Indeed, the only thing the girl does wrong—from a patriarchal perspective—is to survive the attack. However, literary tradition dealt kindly with her. There is no hint of censure in Shakespeare's sources. Gower indeed tells us the king loses all sense of shame ("suche delite he toke therin, / Him thought that it was no sin" [353–54]), while his daughter remains an unwilling victim who "durst him no thinge withseye" (355).

Shakespeare substantially alters this representation of the princess. First he chooses not to dramatize the episode, so that we have no immediate sense of the father's crime. Instead, Shakespeare's Gower relates it in the opening chorus and assigns the girl partial responsibility for the sin. She was, we hear,

> So buxom, blithe, and full of face
> As heaven had lent her all his grace;
> With whom the father liking took,
> And her to incest did provoke.
> Bad child, worse father, to entice his own
> To evil should be done by none;
> But custom what they did begin
> Was with long use accounted no sin.[7]

The princess is now "bad," a willing and active participant in the crime; not the king, but "*they* did begin" the incest. The loss of shame that the *Confessio* ascribes to the father alone, *Pericles* ascribes to them both.

This change involves another. Once the girl's moral and spiritual fall is determined, she becomes herself a figure of carnal temptation to her suitors. She is a temptress. Thus we hear Shakespeare's Gower say, "The beauty of this sinful dame / Made many princes thither frame / To seek her as a bedfellow" (1 Chorus.31–33). In the source versions Apollonius, as far as we are told, never sees the girl. Shakespeare, however, brings her on stage to participate in the riddle scene, where she functions as an emblem of sexual temptation. The Edenic imagery of Pericles's prayer emphasizes her role:

> You gods that made me man, and sway in love,
> That have inflamed desire in my breast
> To taste the fruit of yon celestial tree
> Or die in the adventure, be my helps. (1.1.20–23)

Shakespeare's adaptation of the riddle continues the portrait of the guilty daughter. In the *Historia* the riddle runs: "I am carried with sin, I eat my mother's flesh, I seek my brother, the husband of my mother, son of my wife; I do not find him."[8] As Goolden demonstrates (246–47), the referent of this riddle is Antiochus. It describes both his incestuous relation with his daughter—"I am carried with sin, I eat my mother's flesh"—and his "search" for a son-in-law—"my brother, the husband of my mother, the son of my wife"— whom he cannot find. Gower, the *Gesta,* and Twine all preserve the sense of the original riddle, more or less. Shakespeare is the first to change it radically (Goolden 250–51). His riddle is:

> I am no viper, yet I feed
> On mother's flesh which did me breed.
> I sought a husband, in which labour
> I found that kindness in a father.
> He's father, son, and husband mild;
> I mother, wife and yet his child.

> How they may be, and yet in two
> As you will live, resolve it you. (1.1.65–72)

The "I" of this riddle is exclusively feminine; the referent can only be Antiochus's daughter.[9] Moreover, the terse, cryptic phrasing of the original is replaced by a transparent and fulsome description of the incest. The tone is smug. It is a boast rather than an enigma. While in previous versions Antiochus speaks the riddle, in Shakespeare the hero reads it from a scroll. The king is thus further distanced from the confession it contains. The net effect of the Shakespearean riddle is to shift the responsibility for the incest almost wholly to the princess. Listening to it we forget the chorus telling us that Antiochus provoked his daughter, as we hear the feminine voice of the riddle confessing her active guilt: "I sought a husband, in which labour / I found that kindness in a father." The figure of the girl before us becomes literally emblematic, as the verse interprets her and she illustrates the verse.

Because we hear this "confession," Pericles's initial response—revulsion from the princess, rather than her father—seems natural:

> Fair glass of light, I loved you, and could still,
> Were not this glorious casket stored with ill.
> But I must tell you now my thoughts revolt;
> For he's no man on whom perfections wait
> That, knowing sin within, will touch the gate. (1.1.77–81)

On the level of the plot the heads of the dead suitors, which form the visual background to the scene, are fully explained as the victims of Antiochus's jealous trap. On another level, however, they are clearly emblems of the death the girl embodies: Spiritual death would be the result of involvement with her guilty sexuality. For Pericles the mere knowledge of her sin precipitates his "fall" into the world of experience.

Traditionally, then, the Apollonius legend begins with male aggression, the rape—in effect, the destruction—of a daughter by her father. As Gower puts it, "that was cause of mochel care" (319). Shakespeare changes the ancient romance, subtly but radically, to conform with another ancient legend: Antioch becomes Eden; the victim, the seductress; and her guilty sexuality, the cause of all the woe.

Just as he changes Antiochus's daughter to conform to Christian typology, so too Shakespeare alters the hero's daughter. As one becomes more sinful, the other becomes more saintly—literally saintly, that is: Marina takes on the characteristics of the legendary virgin martyrs.

To some extent the heroine's resemblance to the early saints is part of the *Historia*. Tharsia—Marina's original—conforms to the type of virgin martyr in being wellborn, well-educated, beautiful, chaste and subject to the lustful desires of a high-ranking male. Moreover, like many of the saints discussed above, she descends into a brothel and emerges unstained. The brothel motif is almost peculiar to hagiography, and Kortekaas believes its appearance in the *Historia* reflects a hagiographical source.[10] He notes several verbal echoes of saints' lives, and in particular, of the *Passio Agnetis* (105).

According to legend, the thirteen-year-old Agnes was consigned to a brothel because she refused to sacrifice to the Roman gods. As Caxton tells the story:

> when S. Agnes entered into the bordel anon she found the angel of God ready for to defend her, and environed S. Agnes with a bright clearness in such wise that no man might see her ne come to her. Then made she of the bordel her oratory [. . .]. All they that entered made honour and reverence to the great clearness that they saw about S. Agnes, and came out more devout and more clean than they entered. (De Voragine 2: 248)

Geoffrey Bullough notes the parallels between Tharsia/Marina and Agnes, but dismisses the legend as a source for *Pericles*. "Marina's ordeal in the brothel," he writes, "has something in common with that of early Christian saints, especially Saint Agnes. There is no reason to suppose that Shakespeare thought so of it, but he was obviously eager to get as much pathos and moral beauty out of the situation as possible" (6: 371). I believe, however, that Shakespeare did consciously draw on hagiographical legend and that his chief interest in the situation was not pathos but something we might call "admiration."[11]

The virgin martyrs are characterized by a moral energy that expresses itself in utter defiance of pain and unbridled freedom of speech. (When a wealthy suitor proposes marriage to Agnes, she replies, "Go from me thou fardel of sin, nourishing of evils and morsel of death, and depart" [De Voragine 2: 246].) They are superwomen and every official who unleashes their eloquence has reason to regret it. They are much closer to suffragettes than they are to heroines of melodrama.[12] They are indeed the Church militant.

In the *Historia,* however, in spite of the apparent debt to hagiography, the heroine has little of this formidable energy. She preserves her chastity primarily through her ability to inspire pity. Tharsia throws herself at the feet of each prospective client, weeps, tells her story, begs for pity and is spared (RB 35). Twine, in his account of the brothel episode, remains close to the *Historia,* but heightens the pathos considerably. (The girl declared "her heavie fortune, eftsoones sobbing and bursting out into streames of tears, that for extreme griefe she could scarsly speake" [457]). Again, as in the original narrative, the basis of her appeal is not moral but sentimental, resting heavily on her noble birth ("Take pitty on me, good friend, which am a poor captive, and the daughter of a king, and doe not defile me" [459]). Gower likewise stresses the pity of it:

> But such a grace god hir sent,
> That for the sorowe, whiche she made,
> Was none of hem, which power hade
> To done hir any vilanie. (1436–39)

When the pander's servant is dispatched to rape her, the girl's "wofull pleintes" reduce him to tears (1450–52). Nowhere in the *Historia*, Gower, the *Gesta* or Twine, is there any suggestion that Tharsia presents a moral challenge to the brothel and its inmates.

In contrast to this sentimental tradition, Shakespeare presents a heroine whose power inspires awe.[13] It is literally marvelous. Something indeed of Tharsia's pathos surrounds Marina when we first see her, "weeping for her only mistress' death" (4.1.11), lamenting her fate (4.1.17–20). From the moment of her arrival in Mytilene, however, Marina begins to show her strength. Like the virgin martyrs she is formidably articulate and self-possessed. Like them she issues—both implicitly and explicitly—a moral challenge to all she meets. Like them too, she declares her preference for death to dishonor and like them accepts responsibility for the threat to her chastity (4.2.60–80). The scene concludes with Marina's vow to kill herself before losing her virginity, a vow that is also a prayer (4.2.140–42).

Like Agnes Marina makes "of the bordel her oratory," and would-be patrons emerge "more devout and more clean than they entered." The scene between the "two Gentlemen" (4.5) seems almost an explicit reference to Agnes's miraculous powers of persuasion in the brothel:

> FIRST GENT. But to have divinity preached there!
> Did you ever dream of such a thing?
> SECOND GENT. No, no. Come, I am for no more bawdy houses.
> Shall's go hear the vestals sing?
> FIRST GENT. I'll do anything now that is virtuous,
> but I am out of the road of rutting for ever. (4.5.4–9)

This dialogue outside the brothel forms a farcical prelude to the scene that follows inside. Again the theme is Marina's extraordinary power, and it modulates from the high comedy of the Bawd's outrage ("Fie, fie upon her! She's able to freeze the god Priapus and undo a whole generation" [4.6.3–4]), through the more somber display of the girl's "virginal fencing" (48–57), to the melodrama of her exchange with Lysimachus

(62–114). Unfortunately, the text of this exchange is badly corrupt and probably abbreviated.[14] As it stands, however, it seems clear that Lysimachus undergoes a moral conversion, effected by Marina's words ("I did not think thou couldst have spoke so well, / Ne'er dreamt thou couldst" [99–100]). He leaves blessing Marina and cursing Boult. The violent rebuke he bestows on the latter ("Avaunt, thou damned doorkeeper!" [116]) is again wholly Shakespeare's invention.

Marina faces the final and most serious threat to her chastity when the Bawd sentences her to a summary rape by Boult. Again, the Bawd's anger is a darkly comic tribute to the girl's extraordinary power: "Away with her! Would she had never come within my doors! Marry, hang you! She's born to undo us. Will you not go the way of womenkind?" (145–47). Once again Marina practices her "virginal fencing." She first holds Boult at bay with a riddle,[15] then launches a verbal attack on "the damnèd doorkeeper" (160–64) that reduces him to self-justification and a grudging acquiescence in her plan of escape.

The topos of the brothel thus functions here, as in the hagiographic legends, to illustrate the power of chastity. Subjected to the ordeal of the brothel, Marina is defined as a holy virgin. Her typological significance becomes clearer in the "resurrection" scene, where her role as holy virgin is linked specifically with that of the Blessed Virgin.

Shakespeare was the first interpreter of the Apollonius myth to change the name of the heroine, and the change, as Bradbrook notes, "is not fortuitous in an age which believed that all names were bestowed on creatures by God and discerned by Adam" (189). The name he chose points to her connection with the Virgin. Medieval etymology derived the name Mary from *"stella maris,"* star of the sea" (Warner, *Alone* 262)—an apt description of Shakespeare's sea-born heroine. Marina and Mary are thus closely linked simply by their names. The association extends to their common element, the sea. By the Middle Ages the Virgin had acquired features of a lunar deity: Like Diana—the regnant deity of Shakespeare's play—she was both virgin and patroness of women in childbirth (Warner, *Alone* 155–62, 275–76). She also exercised the moon's sovereignty over waters. As lunar deity, as *"stella maris,"* and *"stella matutina,"* the morning star, the Virgin's relation to mariners was direct and powerful. According to Warner, "the Virgin's association with the sea must never be forgotten, for in a different age the night sky's principal practical function was navigation. Mary's astral character gives her, in medieval legend, hegemony over tempests, not only as the star that leads sailors to safety, but even more directly as a goddess with powers to still the wind and calm the waves" (265).

It is as both "goddess" and guiding light that Petrarch invokes Mary in the ode that crowns his *Rime:*

> Bright Virgin, steadfast in eternity
> Star of this storm-tossed sea,
> Trusted guide of every trustful pilot,
> Turn your thoughts to the terrifying squall
> In which I find myself, alone and rudderless [. . .] [16]

The image is a medieval commonplace—it is the basic metaphor of John of Garland's *Stella Maris*—but Petrarch's language is suggestive in the context of Shakespeare's play. Spiritually "alone and rudderless" on a "storm-tossed sea," Pericles invokes the steadfast virtue of the "bright virgin" who is his daughter ("thou dost look / Like Patience gazing on kings' graves and smiling / Extremity out of act" [5.1.137–39]). Whether Shakespeare knew Petrarch's poem or not, it illuminates some of the associations that surround the potent figure of Marina.[17]

For Pericles *in extremis* Marina has an iconic power:[18] Like Patience on a king's grave or an image of the Virgin, she is an emblem of the transcendent virtue that defeats death, a virtue that he recognizes and to which he pays homage ("Falseness cannot come from thee, for thou lookest / Modest as justice, and thou seemest a palace / For the crowned truth to dwell in. I will believe thee [. . .]" [5.1.120–22]). This quasi-religious response to the girl's *appearance* is absent from the Latin narrative and other source versions. The reverence and wonder reach a climax in the moment of recognition. In the ecstasy of his rebirth Pericles cries out, "O, come hither, / Thou that beget'st him that did thee beget" (5.1.195–96). As Edwards notes, the phrase "was an ancient Christian paradox for the miracle of God the father becoming the son of his virgin daughter" (Shakespeare, *Pericles* 27).[19] Marina's resemblance to Mary is thus explicit. The following lines—"Thou that wast born at sea, buried at Tarsus, / And found at sea again" (197–98)—heighten our sense of the sacred, for they recall the confession of faith known as "The Apostles' Creed": "I believe [. . .] in Jesus Christ his only Son our Lord. Which was conceived by the Holy Ghost, born of the Virgin Mary. Suffered under Pontius Pilate, was crucified, dead, and buried, he descended into hell. The third day he rose again from the dead" (*Book of Common Prayer, 1559* 58). The parallel is not precise, but the rhythm is powerfully present for anyone familiar with the Creed. As Ewbank observes, Pericles' words are a credo ("Word in Theatre" 60). They are a confession of faith in the miraculous facts of his daughter's life: Her birth, death and resurrection.

Responding to the hagiographical element present in his sources, Shakespeare transformed the pathetic Tharsia into his powerfully active heroine, who, like St. Agnes, preaches divinity in the brothel. He named her Marina and drew out the paradox of her spiritual maternity: A virgin, she begets him who begot her. These changes perhaps prompted a third: The transformation of Antiochus's daughter from victim to temptress, answering the ancient paradigm of Eva/Ave. The hero's flight from the guilty sexuality of one leads ultimately to his redemption through the purity of the other.

The Martyred Soldier

Shakespeare's use of hagiography is discreet and there is nothing in *Pericles* to suggest specifically Roman Catholic doctrine. By contrast Henry Shirley's use of hagiography is blatant and *The Martyred Soldier* is politically, if not doctrinally, a Catholic play. The title page of the Quarto[20] claims the play was "sundry times Acted with a generall applause" and there is no reason to doubt this: It has a number of elements likely to ensure a popular success, including martial heroism, bloodthirsty tyrants, supernatural apparitions, music, spectacle, torture, low comedy, high sentiments, chastity imperiled (repeatedly) and a happy ending. Its Catholic bias seems to have eluded the Protestant audience as well as the Master of the Revels, George Buc.[21]

Shirley sets the action in the court of the great Vandal king Genzerick, during the Vandal persecution of the African church in the fifth century.[22] It was an inspired choice for a Catholic playwright.[23] Since the Vandals were heretical Christians, Arians, their persecution of the African Catholics provided an appropriate analogue for Protestant persecution of English Catholics.[23] Shirley protected himself, however, by portraying the Vandals as pagans who worship the Roman deities. The analogue was thus hidden, intelligible only to those whose grasp of late antique history included a knowledge of the Vandals' religion. Most people—perhaps even George Buc—would have missed it. Nevertheless, committed Catholics—like the priests who attended the theaters on day-leave from the prisons[25]—probably would have understood the hidden message. Victor of Vita's contemporary account of the Vandal persecution, Shirley's principal source, had been translated by Ralph Buckland, a Catholic divine, and published secretly in England in 1605.[26] Clearly intended for the edification of the persecuted Catholic community, Buckland's translation highlights the similarities between Africa under the Vandals and England under the Protestants.[27]

Although Victor reports the mass torture of dozens of sacred virgins (59) and the ritual humiliation of matrons (93–94), his narrative lacks the classic virgin-martyr story: Faith, not chastity, is his chief concern. Two climactic moments in his account involve stories of married couples: One partner begs the other to recant, but the Catholic firmly chooses martyrdom.[28] Shirley took up this model of married sanctity and adapted it to conform with traditional hagiography. One of his principal additions was the conventional assault on chastity. His central characters are thus a husband and wife—Bellizarius and Victoria—who are both martyrs: He for his faith alone, she for both faith and chastity.

Shirley also added a secondary love story that permitted a covert allusion to the proposed Spanish match for Prince Charles. Victoria and her daughter Bellina—the only two women in the play—share the functions of the heroine. The matron assumes the spirited role of the traditional female martyr, triumphantly resisting the assaults of the tyrant. The virgin assumes the more passive role of prize: When her chaste love has converted her suitor, the noble Hubert (read: Charles) to Christianity (read: Catholicism), she yields to his honorable desires. In a wholly fictional conclusion, the young couple accept the Vandal throne and the task of generating a Christian dynasty.

Shirley establishes the erotic dynamic of Hubert's conversion early in the play. In response to a bawdy jest about the license for rape during the African war, Hubert tells a friend of a strange adventure:

> after some three hours being in *Carthage*
> I rusht into a Temple, Starr'd all with lights;
> Which with my drawne sword rifling, in a roome
> Hung full of Pictures, drawne so full of sweetnesse
> They struck a reverence in me, found I a woman,
> A Lady all in white; the very Candles
> Took brightnesse from her eyes and those cleare Pearles
> Which in abundance falling on her cheekes
> Gave them a lovely bravery. At my rough entrance
> She shriek'd and kneel'd, and holding up a paire
> Of Ivory fingers begg't that I would not
> (Though I did kill) dishonour her, and told me
> She would pray for me. Never did Christian
> So near come to my heart-strings; I let my Sword
> Fall from me, stood astonish't, and not onely
> Sav'd her my selfe but guarded her from others. (2.1; 191–92)

This is a miniature virgin-martyr story, with a happy ending. Like Marina's, the Lady's chastity is preserved. Unlike Marina, however, the Lady is passive. It is primarily her noble pathos, her tearful beauty, that subdues the savage Hubert. Significantly, she is surrounded by pictures—religious icons, images of the saints—"drawne so full of sweetnesse" they strike "a reverence" in Hubert. The pictures clearly point to the Lady's saintly status: The scene Hubert describes could be represented in one of the paintings. His reverential response to the images anticipates his submission to the Lady (and his later conversion to Christianity by Bellina). Although he has violated the temple, rifling it with his "drawne sword," he refrains from violating her: His sword falls. Shirley mystifies the difference in power between the kneeling, pleading Lady and the advancing, armed soldier. The suppliant woman subdues the conquering man—not, as Marina does, through an inner moral strength but, on the contrary, through a display of weakness. She maintains her chastity because she is so attractively (and virtuously) abject, because she begs for death before dishonor. The key word in this sentimental vision of sanctity is "sweetnesse"; the key image, the suppliant woman.

The paradigm of female power-in-weakness established by Hubert's tale is central to Shirley's representation of virginal chastity. It appears again in the third act when, for no reason inherent in the plot, Bellina enters to kneel weeping before her father and Hubert (whom she has never before met). Shirley clearly wants to display Bellina's pathos and piety for the audience and for Hubert, who is immediately smitten.[29] Although Bellina accepts his marriage proposal on condition of his religious conversion, her vision of their future life together chastens the dangerously erotic bond of marriage:

> In peace Ile sit by thee and read or sing
> Stanzaes of chaste love, of love purifi'd
> From desires drossie blacknesse; nay when our clouds
> Of ignorance are quite vanisht, and that a holy
> Religious knot between us may be tyed,
> *Bellina* here vowes to be *Hubert's* bride:
> Else doe I sweare perpetuall chastity. (3.3; 215)

Thus, even though Bellina is destined for marriage and motherhood rather than martyrdom, Shirley dissociates her explicitly from any hint of carnal passion ("desires drossie blacknesse").

While the virgin Bellina tames Hubert's unruly masculinity with tears and prayers, her mother displays a more vigorous virtue in her defiance of the tyrant Henericke's persecution. Allowed a public audience with her husband (arrested for his Christian faith), she urges Bellizarius to be absolute for death. The enraged king orders Victoria to be bound and raped by one of his camel drivers, before the eyes of Bellizarius and the captive bishop Eugenius. Henericke thus attempts to counter her theatrical defiance with an equally theatrical humiliation, taking vicarious sexual possession of her in an act that would demonstrate his power over her, her husband and her God. The servile status of the assailant obviously adds insult to the injury offered to all three. Victoria's chaste horror of rape ("'tis worse / Than worlds of tortures") elicits the bishop's serene assurance that her spiritual purity will protect her from physical violation:

> Fear not, *Victoria;*
> Be thou a chaste one in thy minde, thy body
> May like a Temple of well tempered steele
> Be batter'd, not demolishe'd. (4.3; 234)

The sexual mythology of the saints' lives is rarely stated so clearly: The truly chaste woman cannot be raped. The subsequent action illustrates both the truth of Eugenius's dictum and the chastity of Victoria's "minde," as two camel drivers and two slaves fail in three successive attempts to rape her at the king's command. The mixture of melodrama and comedy here, as the would-be ravishers suffer ludicrous punishments, recalls the brothel scenes in *Pericles.* Nevertheless, there is a crucial difference in the playwrights' representation of female agency. In spite of her active rhetorical defiance, after the sentence of rape is passed, Victoria is as passive as any heroine of nineteenth-century sensation drama. Her deliverance comes not from within but from heaven.

Frustrated in his attempt to have Victoria violated, the tyrant orders her thrown into a dungeon, chained and starved. Imprisonment and starvation—symbolic death—are common elements in hagiography. So too are angelic intervention and a physical transformation of the saint, who is often healed and/or supernaturally nourished while in prison.[30] All these elements appear in De Voragine's life of Katherine, who is visited by angels while in prison and fed by a dove: Later the emperor "commanded her to be brought tofore him, and when he saw her so shining, whom he supposed to have been tormented by great famine and fasting, and supposed that some had fed her in prison, he was fulfilled with fury and commanded to torment the keepers of the prison [. . .]" (7: 22). Enthralled by Katherine anew, he offers her marriage ("thou shalt triumph as a queen in my realm, in beauty enhanced" [7: 23]), but she rejects the earthly king in favor of the heavenly.

This legendary life of Katherine informs Shirley's fifth act. After three days and three nights, Victoria's jailors are astonished by an angel who *"ascends from the cave, singing"* (5.1; 241). Considerably disturbed, the king is stricken with a fit of necrophilous desire ("My longings feast with her, though her base limbes / Be in a thousand pieces" [243]),[31] when Victoria *"rises out of the cave, white"* (244).[32] Like St. Katherine, she is transfigured, more beautiful than ever, and ready to do battle with the tyrant. In spite of the King's importunity, Victoria refuses to save Bellizarius at the cost of her chastity, and instead cheers him on to martyrdom. Flights of angels—or at least two—sing him to his rest and urge Victoria to follow.

Victoria, however, has to endure a final attempt on her chastity. Initially she rebukes the king's lust in striking terms:

> Thou foole, thou canst not;
> All my mortality is shaken off;
> My heart of flesh and blood is gone; my body
> Is chang'd; this face is not that once was mine.
> I am a Spirit, and no racke of thine
> Can touch me. (248)

For a moment it seems we are to believe that Victoria has passed through death, and that her body, like the risen Christ's, is reconstituted, immortal. However, since the king—who is suddenly transformed into a Volpone-like sensualist—is able to force her onto a "day bed" his minions instantly provide, I think we have to understand her boast metaphorically, not literally. The king wrestles with flesh and blood, not a spirit. However, spiritual help is at hand. The stage directions call for two Angels to enter *"about the bed"* (248). As hagiographic convention demands, Victoria cries out for death ("O that some rocke of Ice / Might fall on me and freeze me into nothing" [248]), and the angels oblige. One or both declare(s), "These Starres must shine no more; soule, flye away. / Tyrant, enjoy but a cold lumpe of clay" (248–49). Briefly, as he hovers over her body, it seems as if Henericke, like the necrophilous tyrants of legend, will do just that. We watch fascinated, awaiting the event. However, Henericke is dismayed to discover that her lips are as cold as winter, or death, to his kiss and promptly consigns her body to destruction ("Since me it hated it shall feele my hate: / Cast her into the fire" [249]). Just as the enraged king has ordered a general massacre of all the Christians, the divine wrath overtakes him. Henericke meets his well-deserved and appropriately melodramatic end as *"A Thunder-bolt strikes him"* (249).

As this brief account demonstrates, *The Martyred Soldier* fully endorses the sexual mythology of the saints' lives: The truly chaste may be assaulted, but not violated. Exploiting the sensational potential of the situation, Shirley dramatizes the miraculous preservation of chastity not once, but four times. While Victoria is at moments heroically active—in her "persuasion" of Bellizarius, in her defiance of Henericke as she rises from her grave-like prison—she is also passive before the initial threat of rape. Faced with Henericke's own lust, Victoria's reply ("Thou foole, thou canst not [. . .]") is spirited but ineffectual, providing the thrill of conflict without the substance. Like the legendary Drusiana, she escapes violation only by voluntary death, a suicide glorified by ministering angels. Her active heroism finally differs little from Bellina's passive piety: Both derive from an idealization of feminine weakness, and both function only within the limits circumscribed by a highly sentimental vision of chastity.

The Virgin Martyr

Even though *The Virgin Martyr*,[33] like *The Martyred Soldier*, dramatizes the life and death of a saint, there is nothing in it to suggest Catholicism: Indeed, recent critics have persuasively interpreted the play as a Protestant allegory of continental religious strife.[34] Evidently prompted by the success of *The Martyred Soldier*, Massinger and Dekker repeated many of its essential elements—martial heroism, bloodthirsty villains, supernatural solicitation, spectacle, violence, low comedy, high sentiments and chastity imperiled— but subjected them to a treatment at once more sophisticated and more sensational. In contrast to Shirley's play, in which Henericke's *pro forma* assaults on Victoria are represented with transparent naivete, *The Virgin Martyr* dramatizes the sexual violence of hagiography with a sustained and pornographic emphasis on the female body as the object of male anger and desire.

The eroticization of the female body begins in 1.1 with the introduction of the beautiful virgins, Caliste and Christeta. These sisters play a minor role in the legend of St. Dorothea, the play's eponymous virgin martyr: As apostate Christians, they were instructed to win Dorothea back to paganism; she reconverted them, and all three were subsequently martyred.[35] Massinger and Dekker use these lesser saints as magnifying doubles for Dorothea, and their role as surrogates is emphasized by their close connection with her: Their parents, we learn, were "neere in love," and the three girls "from the cradle were brought up together" (3.1.70–71). Their father, Theophilus, the chief persecutor of Christians in Antioch, is thus a father figure for Dorothea; his sacrificial murder of his daughters anticipates both the martyrdom of Dorothea and his key role in it. In one version of the legend these girls are Dorothea's sisters (Peterson 18), but in none of the known accounts are they related to Theophilus. This association is apparently the playwrights' innovation, and it is a significant one: It allows them to dramatize the father-daughter tension that is a recurrent feature of saints' lives.[36]

The opening scene establishes the symbolic importance of the sisters. Their entrance, in the company of a priest holding the image of Jupiter (1.1.23.s.d.), is unexplained in terms of the plot. Massinger simply parades them across the stage for their father (and the audience) to look at, emblems of filial and pagan piety. The demonic Harpax interprets their meaning

("Looke on these vestals, / The holy pledges that the gods have giv'n you, / Your chast faire daughters" [23–25]) and rehearses their history for us. Religious allegiance is figured in sexual terms: The girls, "Seduc'd by an imagin'd faith," "quite forsaking / The gentile gods, had yeelded up themselves / To this new found religion" (28–31). Reclaimed by "the authority of a father" (34), they "are now votaries in great *Iupiters* temple, / And by his Priest instructed, growne familiar / With all the Mysteries" (38–40). Given the metaphor of seduction, the "Mysteries" into which the sisters have been initiated suggest erotic pagan rites—a connotation strengthened by the presence of the male Priest (their "instructor") and the adjective "familiar," which carries a sense of sexual intimacy (*OED* 2c). I do not mean to suggest that Caliste and Christete are literally temple prostitutes—on the contrary, they are "vestals"—but rather that the language of this passage would prompt the audience to see them as sexually, as well as spiritually, compromised. This sense of the sisters as "fallen" would grow during their dialogue with Dorothea in 3.1, when the latter implicitly accuses them of "loose lascivious mirth" (56), and declares they have given up immortal happiness for "an Houre / Of pleasure here" (59–60). Overcome with "a vertuous and religious anger," Dorothea castigates the sisters in terms that explicitly link pagan worship with sexual license:

> Your Gods, your temples, brothell houses rather,
> Or wicked actions of the worst of men
> Pursu'd and practis'd, your religious rites,
> O call them rather iugling mysteries,
> The baytes and nets of hell, your soules the prey
> For which the Divell angles [. . .] (108–13)

From the first appearance of Caliste and Christeta in act 1, the female body is thus presented as a contested "patriarchal territory,"[37] a key site in the struggle between paganism and Christianity. Religiously, as well as sexually, a woman may submit to the authority of her father/God, or yield to seduction by another male/god. In either case, in this patriarchal religious and social structure she is significant in terms of her allegiance, rather than of herself. As the governor Sapritius puts it, the girls' "sweet conversion" to paganism expresses "as a mirror" the "zeale and duty" of their father (1.1.117–18). Massinger reinforces the symbolic function of Caliste and Christeta soon after their first appearance when he brings them back onstage for presentation to the emperor and his daughter (1.1.116). Again their religious history is rehearsed, and we hear from Theophilus himself how he, as a judge, watched their "tender limbes" racked, whipped and tortured, but as a father wept (169–98). Again, the "beauteous virgins" are there to be looked at, while the dialogue emphasizes their physical bodies as at once desirable ("they are faire ones, / Exceeding faire ones" [165–66]) and the objects of torture: We are prompted to imagine the "stripes spent on their tender limbs" (186). Now chastened, silent—they do not say a word—and obedient, they are trophies of Theophilus's paternal authority and religious zeal.

The scene of the sisters' martyrdom (3.2) visually echoes their appearances in act 1. Again, they enter in procession, following the priest with the image of Jupiter. The parade is more solemn this time—it includes *"Incense and Censors"* (32.s.d.) and the apparently submissive Dorothea—but again its function is to display the girls as emblems of their father's piety to an onstage audience. Theophilus's exclamation of passionate delight ("Their mother when she bore them [. . .] fild not my longing heart / With so much joy" [29–31]) suggests a quasi-incestuous pleasure in his daughters' performance (they satisfy him more deeply than his wife did). "[R]avisht/ With the excess of joy" (42–43), he now embraces all three girls. Massinger juxtaposes these ecstatic hugs with Theophilus's brutal assault on his daughters after they publicly reject the worship of Jupiter. As Harpax points out, the father's "Honour is ingag'd" (96), and to redeem his honor in the eyes of the watching world, Theophilus slays his daughters, deliberately and religiously:

> Come you accursed, thus by the haire I drag you
> Before this holy altar; thus looke on you
> Less pittifull then Tigres to their prey.
> And thus with mine owne hand I take that life
> Which I gave to you.   *Kils them.* (111–15)

For the audience the spectacle of murder is heightened by the excitements of parricide and sacrilege.

The violence of this scene is also gendered: It is not simply punishment of religious non-conformity, it is an expression of patriarchal ownership, and it is enacted in sexually symbolic terms. The murder weapon—sword or knife—is clearly phallic; however, the image of a man dragging a woman by the hair also connotes sexual aggression.[38] Marston draws on this icongraphy of sexual violence in act 3 of *Sophonisba:* When "SYPHAX, *his dagger twon about her hair, drags in* SOPHONISBA *in her nightgown petticoat"* (3.1.1.s.d.), the stage imagery announces the villain's intentions of rape. Similarly, in Chapman's *Bussy D'Ambois* the jealous Montsurry enters *"unbraced, pulling* TAMYRA *in by the hair"* (5.1.1.s.d.) before he tortures and stabs her in an erotically charged fury.[39] Theophilus's brutal dragging of his daughters by their hair evokes this tradition of male anger and desire. Betrayed by his daughters, who have again been "Seduc'd by an imagin'd faith," Theophilus publicly repossesses them through physical abasement and sacrificial murder.[40] Sensational as this is, there is more to come. Dorothea is reserved for torture, and her page's words of comfort enhance the audience's anticipation of it: "These Martyrs but prepare your glorious fate, / You shall exceed them and not imitate" (133–34).

Dorothea's "glorious fate" entails a scene of spectacular and prolonged sexual threat. As Angelo promised, her martyrdom exceeds that of Caliste and Christete, but it also recalls theirs. Again, the female body is the object of paternal desire and anger. Initially, however, these passions are mediated through the figure of Dorothea's suitor Antoninus. Sick with love, he is close to death when his father, the governor Sapritius, appears at his bedside dragging the girl *"by the Haire"* (4.1.59.s.d.). His language is as violent as his gesture ("Follow me thou damn'd Sorceres, call up thy spirits, / And if they can, now let'em from my hand / Untwine these witching haires" [60–62]). He first urges Antoninus to rape Dorothea ("Breake that enchanted Cave, enter, and rifle / The spoyles thy lust hunts after" [72–73]),[41] then invites the rest of the assembled company to step aside with him and "unseene, be witnesse to this battry" (79). His repeated comments on the action—chiefly expressions of impatience with Antoninus—call attention to the theatrical nature of the assault, and the voyeuristic pleasure it offers the audience. So too does Dorothea's cry ("O guard me Angels, / What Tragedy must begin now?") and Antoninus's answer: "When a Tyger / Leapes into a tymerous heard, with ravenous Iawes / Being hunger-starv'd, what Tragedy then begins?" (65–68). Unable to do his father's bidding, Antoninus relents when Dorothea kneels and begs for death: "O Kill me," she cries, "And heaven will take it as a sacrifice, / But if you play the Ravisher, there is / A Hell to swallow you" (97–100). Dorothea's words once more recall the sacrificial death of Theophilus's daughters and confirm her sanctity: She prefers death to sexual dishonor. She does not explain why, compared to rape, murder is a pious act: She does not need to. She appeals to a value—chastity—that the theater audience recognizes (as Antoninus does) as absolute. This plea is as close as Dorothea gets to action in this scene, and nearly all she says. Like Shirley's Victoria, and unlike Shakespeare's Marina, she is an almost wholly passive victim of sexual humiliation and threat.

As Peterson notes, this episode owes something—principally the motif of the lovesick wooer—to the Agnes legend (89). However, it may also owe a good deal to *The Martyred Soldier*. Here, as in Shirley's third act, sexual assault is set up as a theatrical event. Here too, the rape becomes a matter of pride, a public demonstration of power for an authority figure. When Antoninus relents, his father bursts in and castigates him for his weakness. Just as Henericke attempts to punish Bellizarius by forcing him to watch a camel-driver rape Victoria, Sapritius attempts to punish his son:

> thou shalt curse
> Thy dalliance, and heere before her eyes
> Teare thy flesh in peeces, when a slave
> In hot lust bathes himselfe, and gluts those pleasures
> Thy nicenesse durst not touch. (118–22)

The threatened rape is thus a function of the father's rivalry with his son. There is also a strong sense of Sapritius's own erotic attraction to Dorothea: His eagerness for the rape, his prurient desire to watch it, his anger with his son's reticence and his hatred of Dorothea are all insufficiently explained by a paternal care for Antoninus's health. He wants to rape her by proxy. This quasi-incestuous desire again recalls Theophilus's parricidal violation of his daughters.

Like Henericke, though not so comically, Sapritius is frustrated in his efforts. The slave who appears, a Briton, scornfully refuses the office and is carried off for torture. Dorothea points out the moral of this and all such hagiographic episodes: "That power supernall on whom waites my soule, / Is Captaine ore my chastity" (161–62). Like Henericke Sapritius is enraged by his frustration, but unlike him responds with a direct physical attack on his victim. "Plagues light on her and thee," he cries to his son:

> thus downe I throw
> Thy Harlot thus bi'th haire, naile her to the earth,
> Call in ten slaves, let every one discover
> What lust desires, and surfet heere his fill,
> Call in ten slaves. (164–68)

It is, curiously, Dorothea's page Angelo who answers the enraged governor. "They are come sir at your call" (68), he replies, though there is no direction for their entrance. Instead Sapritius falls down apparently dead. I think we have to assume that the servants announced by Angelo are heavenly spirits who smite the offending Sapritius. Once again heaven protects the truly chaste from violation.

Although Sapritius recovers when Dorothea prays for him, he accuses her of witchcraft: "Wheres the *Lamia* / That teares my entrailes? / I'me bewitch'd, seize on her" (181–82). Dorothea's meek response ("I'me heere, do what you please") elicits only further violence ("Spurne her too'th barre"

[181–83]). As Sapritius's final words make clear—"Kicke harder, goe out witch" (185)—the directive to "spurne her" must be understood literally.

Dorothea's torture follows quickly, preceded only by a brief dialogue between Harpax and Dorothea's comic servants, Hircius and Spungius. As their names suggest, they are figures of carnal appetite—goatish lechery (Hircius) and drunkenness (Spungius). Harpax persuades them, Judas-like, to accept the task of beating their mistress. Their bawdy puns establish the attack as another form of sexual assault:

> HIRICUS. The first thing I doe Ile take her ore the lips.
> SPUNGIUS. And I the hips, we may strike any where?
> HARPAX. Yes, any where.
> HIRICUS. Then I know where ile hit her. [. . .] Ile come upon her
>   with rounce, robble-
> hobble, and thwicke thwacke thirlery bouncing. (52–61)

The detailed stage directions for the actual beating suggest the importance of the visual element: *Enter* Dorothea *lead Prisoner, a Guard attending, a Hangman with Cords in some ugly shape, sets up a Pillar in the middle of the stage,* Sapritius *and* Theophilus *sit,* Angelo *by her* (4.2.61.s.d.). The apparatus of torture—the cords "in some ugly shape," the pillar—function as both practical stage properties and symbols of violence. Tied to the pillar at center-stage, Dorothea is surrounded by hostile men, the object of verbal and physical abuse. Sapritius and Theophilus, seated at some distance, are both actors and audience in this theater of torture. She is struck repeatedly, and in the face. Angelo, however, *"kneeling holds her fast"* (85.s.d.), and in spite of Theophilus's injunction to "Beate out her braines" (85), Dorothea's tormentors are unable to harm her. Joyfully, she urges them to "strike home" and "feast" their fury full (92–93). The directions call for Theophilus to leave his seat at this point (93.s.d.). He must approach Dorothea and stare at her, marveling, because he cries to Sapritius:

> see my Lord, her face
> Has more bewitching beauty then before,
> Prowd whore: it smiles, cannot an eye start out
> With these. (95–98)

Theophilus's response suggests that it is Dorothea's sexuality rather than her religion that he hates. Provokingly unhurt, she is also provokingly beautiful. Her smile, her invitation to "strike home," suggest, moreover, a perverse pleasure in the violence she suffers.[42]

Dorothea is led offstage and 32 lines later led back on for her execution. Practically this allows for a change of scene: Presumably the pillar would be removed before the *"scaffold thrust forth"* (4.3.4.s.d.). However the procession on and off also heightens our consciousness of Dorothea as prisoner, as beautiful victim. (In the course of the play she is led onstage three times, once dragged on by the hair, and four times led off under guard.) The interlude also permits a brief dialogue between the sick Antoninus and his friend, Macrinus, that prepares us for the spectacle of the execution. In spite of Macrinus's forebodings, Antoninus is determined to be "an eye witnesse / Of her last Tragicke scene" (4.3.21–22), and the sentimental tone of the scene is set by Antoninus's maudlin grief. When Dorothea enters, under guard, and mounts the scaffold, Antoninus interprets the significance of the action, cuing our response ("See she comes, / How sweet her innocence appeares [. . .]" [32–44]).

The play's sentimentality reaches new heights, however, when Angelo enters shortly afterwards, *"in the Angels habit"*—that is, transfigured and visible only to Dorothea, the demon Harpax and the audience. Angelo now reveals to Dorothea that he had been sent by God to try her charity. With an eye on the audience he declares the moral of her story ("Learne all / By your example to looke on the poore / With gentle eyes [. . .]" [140–42]). Dorothea's final words proclaim her triumphant self-definition as virgin martyr:

> Hereafter when my story shall be read,
> As they were present now, the hearers shall
> Say this of *Dorothea* with wet eyes,
> She liv'd a virgin, and a virgin dies.
>   *Her head strucke off.* (176–79)

This *non sequitur* exposes the play's covert agenda: Dorothea's sexual assault, torture and death have, in one sense, nothing to do with her faith and everything to do with her sexuality.[43] When Dorothea reappears in the last scene of the play, she comes in glory, *"in a white robe, crownes upon her robe, a Crowne upon her head, lead in by* [Angelo] *the Angell,"* attended by the spirits of Antoninus, Caliste and Christeta, also *"all in white, but lesse glorious"* (5.2.219.s.d.).

Threatened with rape, Dorothea is, like Victoria, primarily a passive victim. However, she—like her surrogates, Caliste and Christeta—is subjected to considerably more physical violence than Victoria: Dragged by her hair, thrown to the ground (again by the hair), kicked offstage, tied to a pillar and beaten, and finally beheaded. Verbally too, the violence is greater in *The Virgin Martyr*, while the sexual threat to the heroine (4.1), unrelieved by the comic elements present in *Pericles* and *The Martyred Soldier*, lasts much longer. Throughout Dorothea is presented with a covert eroticism at odds with the play's overt morality. Massinger and Dekker thus offer their audience the subliminal satisfactions of sadomasochistic pornography sanctified by a sentimental piety.

The Two Noble Ladies

By contrast with *The Virgin Martyr*, *The Two Noble Ladies and the Converted Conjuror* has all the freshness of naive art.[44] On the basis of her textual study, Rhoads suggests that the author was amateur, with a "decidedly literary" hand (*Two Noble Ladies* v-vi). If so, one may imagine him a well-educated Catholic, or crypto-Catholic, encouraged by the precedents set by Shirley, Massinger and Dekker to try his own hand at a saint's play. The result was, like *The Martyred Soldier*, a hybrid: A "Trage-comicall Historie"[45] in which hagiographic romance is grafted onto heroic drama.

The play's principal hagiographic debt is to the story of Justina and Cyprian. According to legend Justina was a Christian virgin of Antioch pursued by a pagan suitor. Rejected by Justina, the suitor enlists the help of a magician, Cyprian. When Cyprian sees Justina he himself falls in love and attempts to secure her submission magically. A succession of progressively more powerful devils tries to win the girl to Cyprian's will. Finally the Devil himself admits to the exasperated magician that he is powerless to sway Justina, protected as she is by Christ. Convinced of Christ's superior power, Cyprian burns his magic books and becomes a Christian. Both he and Justina are subsequently martyred for their faith.[46]

The playwright reworks this contest between demonic magic and Christian chastity in the context of a fictional conflict between the rival states of Antioch, Egypt and Babylon. He both ennobles Justina—she becomes a royal princess and heir apparent to the throne of Antioch—and weakens her. While the legendary Justina, true to type, is resourceful, resilient and defiant, fearing neither flesh nor spirit, the latter-day saint is a Spenserian damsel-in-distress, perpetually in need of rescue. By the time Cyprian begins his attempt on her chastity in act 4, she has already endured threats of death (1.1, 3.3) and of a fate worse than death (1.4)—

gang rape by a lustful band of soldiers—as well as verbal and physical abuse by a pagan tyrant (3.1). From these straits Justina has been rescued variously by a faithful retainer, a mysterious knight (the play's second "Noble Lady," here—like Britomart—disguised as a man and ready to help chastity imperiled) and the magician Cyprian. She has also agreed to marry the prince of Babylon to convert him to Christianity ("To winne a soule to heav'n by yeilding love / may move a virgin hart that has not vow'd /secluded chastitie" [1.4.329–31]).

Cyprian woos Justina, as Volpone woos Celia, with the offer of wealth, power and sensual gratification (4.5.1591–99), and his words echo Volpone's fantastic proposals. Justina, like Celia, responds to this temptation in religious terms:

> Should these vanities
> (faithlesse as are your wondrous promises)
> lead me into the hazard of my soule
> And losse of such ay-lasting happinesse.
> as all earthes glories are but shaddows to? (1600–4)

When Cyprian resorts, like Comus, to an argument from nature, Justina will have none of it ("How ere you please to stile / a lustfull appetite, it takes not mee" [1619–20]). Instead she announces her religious commitment to virginity: "Heav'n has my vow, my life shall never bee / elder then my unstain'd virginitie" (1621–22). This is a sudden development: At the beginning of the scene Justina had declared her religious commitment to her betrothed ("I am the troth-plight wife of Clitophon / the prince of Babilon, hee has my hart" [1576–77]). It is as if, under the pressure of Cyprian's temptation and as the climax of her trial approaches, the romance heroine evolves—or reverts—into the virgin martyr.

Justina's exit leaves Cyprian resolving to use demonic force to subvert her will: "[W]ith greater heat she shall desire her rape/ then I have done. Hells hookes she cannot scape" (1640–41). He is determined to arouse her desire. One of the strategies listed by Cyprian is to "Sing in her eares the scapes of Iupiter [. . .] / That all her senses may at once enforce / a carnall eagernesse to be enjoy'd" (1740–48). By implication, stories of mythological rape ("the scapes of Iupiter") are presumed to function as an erotic stimulus for women, who will enjoy identification with the beautiful victim. Cyprian here enunciates the same logic that animates Volpone's invitation to Celia to "act Ovid's tales": "Thou like Europa [. . .] and I like Jove" (3.7.220–21). Significantly Cyprian speaks of the sensuality he hopes to arouse as a desire *"to be enjoy'd"*: Justina will not actively take her own pleasure in his body, but rather submit her body to his pleasure.

We remain in suspense, waiting for the outcome of Cyprian's threat through the interval between acts 4 and 5. It is not until the second scene of act 5 that Cyprian reappears and we learn his magic has so far proved unavailing. Cantharides, his demonic aide, confesses that all attempts to rouse Justina's sensual nature have failed. Enraged, Cyprian demands: "Let me see / this Christian Saint which I (in spite of hell) / am forc'd to worship" (5.2.1750–51). In response *"Iustina is discovered in a chaire asleep, in her hands a prayer book, divells about her"* (1752–54). For the next forty lines Justina remains asleep, surrounded by demons, while the magician hovers over her. The threat of rape is explicit: Cantharides is at Cyprian's elbow to urge force (1764–65). Cyprian is too much of a hedonist to comply instantly, however, and he protests:

> That abates of pleasures sweetnesse; if such violence
> must be the end yet the beginning shall
> be milde, and I will steale into my roughnesse
> by soft gradations. let sweet musicke plead
> with ravishing notes to winne her maidenhead.
> *Musick. A song.* (1766–71)

The "ravishing notes" of the song are followed by Cyprian's own breathless hymn to Justina's beauty as he gazes at the sleeping woman (1776–88).

There is no precedent for this scene in *The Golden Legend,* probably the most detailed account of the story available to the playwright. Neither Cyprian nor his rival enjoy this access to Justina in any version I have found. (In *The Golden Legend* the devil, disguised as "a fair young man" does get into her bed, but Justina routs him with the sign of the cross [De Voragine 7: 168]). However, this scene recalls several others in early modern drama in which a reclining, unconscious woman is the object of a desiring male gaze. Perhaps most directly it recalls the magical assault of Panthea in *The Wars of Cyrus*[47] where the would-be seducer, Araspas, and a magician attempt to charm the sleeping heroine. Her virtue proves too strong, and when she wakes she angrily repels her assailant (3.2.764–853). Cyprian's voyeurism may also owe something to Tarquin's predatory survey of the sleeping Lucrece in Shakespeare's poem and Heywood's play; to Othello bending murderously and lovingly over the sleeping Desdemona;[48] to Cerimon and his assistant standing over the unconscious Thaisa in *Pericles;*[49] to Iachimo poring over Imogen as she sleeps; and, in *The Second Maiden's Tragedy,* to the Tyrant's gloating over the dead Lady, who is, like Justina, also reclining in a chair.[50]

The peculiar erotic power of these scenes lies primarily in the disparity between the observer and the observed. The woman lies entirely passive and vulnerable; the male gaze possesses her body unchallenged.[51] Members of the audience may be excited and/or disturbed, depending on the extent to which each identifies with the male predator or female victim.[52] Our response is not simply a matter of suspense ("what happens next?"), though that is obviously a large part of it. We are also excited or disturbed by the violation of the victim that is enacted by the male gaze, marked rhetorically by the poetic blazon of the female body and signified emblematically by the positions of the man (erect, dominant) and the woman (supine, passive).[53] Given the residual early modern belief in "overlooking"(*OED* 7)—that is, the ability to cause harm by looking—the perceived disparity in power between the male subject and the female object of the gaze would be even greater for Jacobean audiences.[54]

Compared to Iachimo's inspection of Imogen, however, Cyprian's survey of Justina is innocuous. Sensual detail scarcely intrudes. His response is more reverential than lascivious. Thus, while this scene recalls other, more intense, erotic moments, the playwright is primarily interested in dramatizing the marvelous power exercised by the saint, in all her weakness, rather than in exploiting the erotic potential of the situation. Like Shirley, he treats chastity sentimentally, and mystifies the actual disparity in power between the heroine and her potential ravisher. When Cyprian attempts to kiss Justina *"she starts, wakes, and falls on her knees,"* crying "Forbid it heav'n" (5.2.1792–93). At this most critical moment Justina is pathetic rather than, like her prototype, formidable. Like Henry Shirley's saintly women, she kneels to implore protection, and her trust is vindicated. According to the directions, Justina *"looks in her booke, and the Spirits fly from her"* (1796–97). It is precisely the heroine's weakness that the playwright emphasizes. Again and again Cyprian alludes to it: "[H]as this weake woman pow'r to make hell shake?" (1798); his powers are "sham'd with womans learning!" (1814); he is disarmed "by a weake woman's ffaith" (1817). Awed, Cyprian kneels to her (1817–18) in a gesture that recalls Hubert's submission to the kneeling Bellina in *The Martyred Soldier.* Finally Cyprian concludes,

> If over this weake peece hell have no pow'r,
> then there's a skill can make the weakest man
> more potent then the strongest feind of hell;
> and that shall be my studdy. (1835–38)

Confronted by Justina's faith, the "feinds" that threaten Cyprian *"roare and fly back"* (1847). A *"patriarch-like Angell"* enters to further confound the devils—they *"sinck roaring"* (1860)—convert Cyprian's "carnall lust" to "love of heaven" (1869–70) and foretell the martyrdoms of both saints. Cyprian rejoices in his fortunate fall:

And blessed be Iustina, and the day
I first did see thee: yea my very lust
deserv's a blessed memorie, since that
was the first, though a foule step to this blisse. (1893–96)

Justina's victimization is thus sanctified; her subsidiary role as object of desire in the drama of Cyprian's conversion appears divinely ordained.

In addition to the story of Justina and Cyprian, *The Two Noble Ladies* dramatizes a second hagiographical theme: The persecution of a daughter by her incestuous father. As I have observed, many of the virgin martyrs are persecuted by their fathers as well as sexually threatened by a lustful official. In the legend of the Irish saint Dympna the paternal and lustful figures are united. Dympna was a princess whose father, burning with incestuous desire, "began to wooe her with flattering and faire promises, offering her all glorie, riches and honour, if she would be his wife" (*Lives of Women Saints* 43). When she refuses he angrily threatens to force her. Dympna manages to escape with a priest, but her father pursues and kills her. (The difference between Dympna and the daughter of Antiochus in the Apollonius legend is thus simply that the latter is raped by her father and the former murdered by her father for resisting rape. Again, the price of sanctity is chastity unto death.)

Miranda (the second of the play's "Noble Ladies") is a Dympna figure. In the first act her father, the Souldan of Egypt, suddenly and publicly begins to court her. In its formality the scene recalls the opening of *The Second Maiden's Tragedy:* Here, as there, the tyrant's offer is rather a statement of intent to marry the heroine than an earnest wooing. Here, as there, the tyrant's offer of marriage is a public display of power and the daughter's response a public display of virtue. Here again the heroine displays that virtue by firmly resisting the temptation of "all glorie, riches, and honour." Unlike the tyrant in the earlier play, however, the Souldan passes quickly to threats of coercion. Miranda's ability to best him—and his flattering courtier, Colactus—rhetorically, demonstrates her superior moral strength:

> SOULDAN. To bee the Queen of Egypt, is this shame?
> MIRANDA. To glorifie with incest, is this honour?
> SOULDAN. Thou shalt agree *Miranda,* we must wed.
> MIRANDA. Agree with death, not with a fathers bed.
> COLACTUS. Beware faire princesse to displease your father.
> MIRANDA. Beware great father to displease the gods. (1.2.180–85)

Thus, even though Miranda is not a Christian, she behaves like a Christian saint, preferring death to dishonor and advancing religious objections to her father's suit. Like Dympna too, Miranda manages to escape, though alone and disguised as a man. As she tells the audience, "The care of honour / (I being manlike) chang'd mee thus to man" (1.4.360–61). By implication, Miranda's heroic care of honor—that is, chastity—qualifies her for the adjective "man-like," and renders her unlike other women. Her disguise is no disguise: On the contrary, it reveals her superior, masculine, nature. Like Dympna's father, Miranda's pursues her vengefully. In his rage his language reaches a level of brutality unparalleled in the previous plays. He out-Herods Herod, threatening "to mangle her enticing face, / seare up her tempting breasts, teare wide her mouth, / and slit her nose" (2.1.553–55). Unlike the violence with which Dorothea is threatened, however, this hideous disfigurement is never effected. It remains a threat, hovering over the disguised heroine.

Justina and Miranda thus share the role of virgin martyr: Justina in a passive, "feminine" mode, Miranda in an active, "masculine" mode. Justina's weakness allows Cyprian a demonstration of divine strength, and the audience a theatrically exciting scene of sexual danger. Miranda's masculine valor makes her the exception that proves the rule of female nature. While her heroic activity presents a welcome relief from Justina's timorous passivity, the contrast is more apparent than real. Like Pamela and Philoclea, the paired heroines of Sidney's *Arcadia,* the two noble ladies offer alternative versions of the same chaste ideal: Both affirm their worth by suffering for the sexual and political aggression of men. Both are beautiful victims.

The Jacobean "saint's plays" examined in this chapter—*Pericles, The Martyred Soldier, The Virgin Martyr* and *The Two Noble Ladies*—amply dramatize the sexual mythology of Christian hagiography. They represent sexual assault as a pressing invitation to sensual pleasure that the virtuous heroines refuse, an invitation provoked primarily by female beauty. Though Victoria and Dorothea die, all of these latter-day saints prove the hagiographic rule of the chaste woman's vaginal inviolability. In the words of St. Ambrose, "The virgin of Christ can be exposed to shame, but not contaminated"[55]; Marina, Victoria, Dorothea, Justina and Miranda are, indeed, exposed to shame, but none is contaminated by rape. Of the five, Marina is perhaps threatened with the greatest "shame"—prostitution in a brothel—and demonstrates the greatest inner strength in response to her danger: She talks her way out. Victoria, Dorothea and Justina, however, are preserved miraculously by external powers. Their various providential rescues resemble the episodes described by Basil of Ancyra:

> In the time of the persecutions virgins who had been pursued because they remained faithful to the Husband and were given to impious men managed to preserve their bodies from being defiled; for He for whom they had suffered paralyzed and struck with impotence the violence those impious men tried to use on their bodies [. . .]. (qtd. in Rouselle 192)

In *The Martyred Soldier, The Virgin Martyr* and *The Two Noble Ladies* the scenes of sexual threat serve an obvious doctrinal purpose (divine strength protects feminine weakness) and provide theatrical excitement. In *The Virgin Martyr* the sexual threat also includes an element of erotic entertainment for the theater audience. Finally, each of the plays includes the motif of a wooer's conversion: Lysimachus *(Pericles),* Hubert *(The Martyred Soldier),* Antoninus *(The Virgin Martyr),* Cyprian and Clitophon *(The Two Noble Ladies)* are all spiritually "converted" by the chastening power of the virgins they desire. In the last play this motif becomes an explicit rationalization for sexual assault, as Cyprian rejoices in his fortunate fall ("my very lust / deserv's a blessed memorie" [5.2.1894–95]). This rationalization of assault, latent in the hagiographic sources for these plays, becomes thematically prominent in the marriage plays explored below in chapter five.

Chapter 3 🏵

The Classical Paradigm: Lucrece and Virginia

> And to a woman, what stoare of examples are there to instructe her in her dutie, eyther for the maried to kepe her fayth to her husband, or the unmaried to defend her virginitye, with Virginya [. . .].

> —Geoffrey Fenton, *Tragical Discourses,* on the importance of reading Livy

The legend of Virginia, which Geoffrey Fenton recommends so earnestly to his female readers, has lapsed into obscurity. In early modern England, however, it enjoyed wide currency among a literate elite, trained in humanism. Closely related to the story of Lucretia (Ogilvie 477), it evokes the same political paradigm of rape as does its more famous analogue: The death of a sexually threatened/violated female becomes instrumental in liberating her community. In contrast to the martial heroism of the avenging males, the role of the female is primarily sacrificial. As an innocent victim she absorbs the evils of political oppression—expressed sexually—and pays for them with her death. This narrative of rape and liberation appeared again and again on the Jacobean stage, both in the direct dramatizations of the classical legends and in a variety of transformations. In this chapter I look briefly at some Elizabethan versions of the legends[1]—R. B.'s *Apius and Virginia* (c. 1564), Shakespeare's influential *Lucrece* (1594), and *Titus Andronicus* (c. 1591)[2]—before turning to the two Jacobean plays that dramatize the classical pattern directly: Heywood's *The Rape of Lucrece* (1607) and Webster's *Appius and Virginia.*[3] I begin, however, with that favorite Renaissance text, Livy's *History.*

The story of Lucretia received its definitive shape in Livy's history of Rome, written around 27 BCE, at the end of nearly 70 years of anarchy and civil war. Turning from what seemed to him the corruption and confusion of his own time to the solace of the legendary past, Livy exhorted his readers to consider the antique virtue that had made the city great.[4] His first book culminates in the drama of Lucretia, the story of how Rome shook off Etruscan domination and the monarchy (bk. 1, ch. 57–60). One night, while stationed with the army, the prince—Sextus Tarquin—and his companions discuss the excellence of their respective wives. They agree to test the women by a surprise visit: Collatinus wins the wager because his wife, Lucretia, alone of all her sex, is virtuously employed spinning. However, this encounter awakens Tarquin's lust. He returns to the house in secret and rapes her at sword point. (Lucretia "yields" to Tarquin when he threatens to leave the body of a slave beside hers and tell the world that he killed them in the act of adultery: The threat is clearly included in the story to explain Tarquin's success—his drawn sword, superior strength and the circumstances of the attack are by themselves insufficient. Had he simply threatened death, the Roman matron, like the Christian martyrs, would have been expected to die resisting.) Summoning her father and husband, Lucretia asks them to avenge her, reveals the identity of the ravisher and kills herself before their eyes. Appalled, the men swear an oath of vengeance, and, led by Lucius Junius Brutus, rouse the populace. In a series of battles, the Romans eventually expel the Tarquins forever.

Livy's third book details Rome's preservation of its freedom; at its center is the drama of Virginia, and the plebian revolt against the patrician decemvirs (bk. 3, ch. 44–54). Virginius, a plebian and a centurion on active duty in the army, has betrothed his daughter, Virginia, to another plebian, Icilius. The chief of the decemvirs, Appius, falls in love with the beautiful girl and conceives a plan to kidnap Virginia. His agent, Clodius, claims that she is really the daughter of his slave, and thus legally his property. Appius himself presides as judge in the case. Ordered by the tyrant to surrender his daughter, Virginius instead seizes a knife and slays her, as he says, to preserve her freedom. He then flees to the army and raises a mutiny. Led by Icilius and Virginia's uncle, Numitorius, the people of Rome join the rebellious army, and the decemvirs are overthrown. Livy's carefully wrought narration of these episodes inspired countless retellings.[5]

For early modern England as for post-Republican Rome the legends of Lucretia and Virginia provided a golden world of heroic virtue. In spite of the

Christian prohibition of suicide, Lucretia was commonly cited as a model of female excellence. Virginia's murder presented a greater problem for a Christian audience, but she too enjoyed the status of exemplar. Their Roman *pudicitia* became Christian chastity, and both were invested with an aura of sainthood.[6]

The stories of Lucretia and Virginia thus offered multiple attractions for the playgoing public. They combined all the appeal of the hagiographical legends examined in chapter 1—exemplary heroines, melodrama, sexual violence—with no taint of Roman Catholic "superstition." Perhaps more important, they possessed the advantage of male heroism. In the stories of the virgin martyrs men play primarily negative roles—persecuting suitors, fathers, provosts, tyrants, torturers.[7] By contrast, in this classical paradigm male valor triumphs over tyranny. At its close a new political order is established in which male lust and aggression have been expelled and male property rights vindicated.

Elizabethan Versions

Although primitive, R. B.'s "tragicall comedie" exemplifies the Christian reinterpretation of Virginia's story. The title page promises "a rare example of the vertue of Chastitie, by Virginias constancy, in wishing rather to be slaine at her owne Fathers handes, then to be deflowred of the wicked Judge Apius."[8] The advertisement illustrates the primary difference between the Renaissance versions and their classical source. In Livy's account Virginia is a silent victim: Her father stabs the girl suddenly before Appius can seize her. We hear nothing of her feelings before she dies. By implication, however, Virginia, like everyone else, is taken by surprise. After the killing, Virginius escapes; Icilius and Numitorius astutely display the body to the populace.

In Chaucer's "Physician's Tale"—R. B.'s immediate source—the death scene is considerably extended.[9] Virginius still takes the initiative, but he gives the girl a choice of "outher deeth or shame" (214). After fainting with horror this "gemme of chastitee" (223) bravely and deliberately chooses death. Her father then cuts off her head and takes it to Appius. Like many other details of Chaucer's narrative, the decapitation recalls the hagiographical accounts of the virgin martyrs.

In R. B.'s version Virginia takes the initiative, dissuading her father from suicide and, like a Christian martyr, eagerly pleading for death:

> Thou knowest, O my father, if I be once spotted,
> My name and my kindred then forth wilbe blotted;
> And if thou my father should die for my cause
> The world would accompt me as gilty in cause.

> Then rather, deare father, if it be thy pleasure,
> Graunt me the death; then keepe I my treasure,
> My lampe, my light, my life undefiled:
> And so may Iudge Apius of flesh be begiled.
> Thus upon my knees with humble beheste,
> Graunt me, O father, my instant requeste. (794–803)

Virginia, moreover, urges her father to take her head to Appius on his knife, and "Bid him imbrue his bloudy handes in [her] giltles bloud" (821–22). Virginius does so and makes a spectacular entrance into the courtroom ("Ah wicked Iudge, the virgin chaste / Hath sent her beutious face / In recompense of lechour gaine . . ." [883–85]). Virginia is thus heroically responsible for her own death and the subsequent deployment of her body.

In *Titus Andronicus*—Shakespeare's "Virginia" play—the heroine's rape and death have a political rather than a religious significance.[10] Like her prototype, Lavinia is a symbol of Rome itself—a metonymic relationship established in the first act when she, like Rome, is subject to the rival claims of Bassianus and Saturninus. (Bassianus here corresponds to Icilius, Virginia's betrothed, Saturninus to the predatory Appius.) Shakespeare varies the classical pattern by having the girl's father favor the tyrant figure: Titus unwisely awards Lavinia, as he awarded the crown, to the elder brother (1.1.244), though the rest of the Andronici actively support Bassianus's seizure of Lavinia. Lavinia's own role in this scene—the silent object of a property dispute—foreshadows her subsequent fate. When Titus demands that Lucius restore her to the emperor, his son responds, "Dead, if you will, but not to be his wife, / That is another's lawful promised love" (297–98).[11] This is, in effect, Virginius's answer to Appius.

Shakespeare treats Lavinia's rape and mutilation with a shocking realism, unequaled by any representation of sexual assault on the Jacobean stage. There is nothing erotic about it, nor is there anything heroic about the rapists: Chiron and Demetrius are brutal thugs.[12] The chief villain of the rape scene, however, is Tamora—Lavinia's opposite—who, in spite of the girl's pleas for "present death," inexorably sentences her to a fate worse than death (2.3.173–80). In her anguish Lavinia begs to be thrown dead "into some loathsome pit," free from the violation of "man's eye" (176–77). Exactly the opposite happens to her: Raped and mutilated, she becomes a spectacle, the object of every man's eye. Marcus shows her to Titus declaring, "This was thy daughter" (3.1.63). In response Lucius cries "Ay me! This object kills me" (64), and Titus asks Lavinia, "what accursèd hand / Hath made thee handless in thy father's sight?" (66–67). Titus's question signals his appropriation of his daughter's suffering. Her wounds are primarily significant as an image of *his* grief, a map of *his* woe. Lavinia's chief function in act 3 is to evoke Titus's emotions: His pity, his grief, his anger. It is primarily through Lavinia that Titus becomes tragic.[13]

At the height of his suffering Titus asks when his "fearful slumber" will end (3.1.251), and Marcus answers:

> Thou dost not slumber: see thy two sons' heads,
> Thy warlike hand, thy mangled daughter here;
> Thy other banished son with this dear sight
> Struck pale and bloodless, and thy brother, I,
> Even like a stony image, cold and numb. (253–57)

In Marcus's summary Lavinia is grouped with the two heads and the hand: Though living, she is like a severed limb or head, grotesque. Indeed, Marcus implicitly associates the ravished Lavinia—whose sight turns him to "a stony image"—with Medusa, who was transformed after her rape into a Gorgon whose visage turned viewers to stone (Ovid, *Metamorphoses* 4.968–79). The monstrous aspect of Lavinia, only suggested here, emerges vividly in the "discovery" scene (4.1) when she chases the child Lucius around the stage: She has become a figure of terror to him. She is no less grotesque gesturing frantically (8–17, 37), "tossing" the pages of Ovid (41), or finally guiding with her stumps the staff she holds in her mouth (75.s.d.). Lavinia's obvious association with Philomela, who exacted such terrible revenge for her rape and mutilation, strengthens this demonic aspect of Shakespeare's heroine.[14]

When Lavinia enters for the final banquet, however, she has "*a veil over her face*" (5.3.25.s.d.). The veil provides a cloak—more symbolic than real—for her deformity and bestows on her the privacy she chastely begged of Tamora. Just as it conceals Lavinia's shame, so Titus, in killing her, metaphorically blots out the stain of her rape: "Die, die, Lavinia, and thy shame with thee, / And with thy shame thy father's sorrow die" (45–46). Titus's words underline his appropriation of her pain: *Her* shame is *his* sorrow and he has had enough. Unlike R. B.'s Virginia, or Chaucer's, Shakespeare's heroine cannot vocally ask for her death. There is no sign in the text to tell us if she is a willing victim or not. In her death as in her passion Lavinia is mute.[15]

In contrast to the political focus of *Titus Andronicus,* Shakespeare's *Lucrece* deals primarily with the subjective experience of the rapist and the victim immediately before and after the crime.[16] The poem is distinguished from previous treatments of the theme not only by its length but also by its eroticism. Shakespeare effectively draws out, as none of his sources do,[17] the time between Tarquin's decision to rape Lucrece and the accomplishment of the crime. He thus exploits to the fullest the dramatic irony arising from our knowledge of Tarquin's intent and Lucrece's ignorance. The thrill of horror we experience as we watch the villain stalking his prey reaches its height as he gazes on his sleeping victim (372–441).

The poem's erotic appeal lies not only in this narrative tactic, however. It is present in the descriptions of Lucrece, especially as she lies asleep. None of the sources focus on the sleeping Lucrece or use the opportunity for a graphic description of her body. They confine themselves to Tarquin's threat and Lucrece's fear. By contrast Shakespeare emphasizes the role of Tarquin's sight here. As Kahn observes, "[t]he insistently visual coding of Lucrece in the first 575 lines as one who is seen but does not see, who is seen but not heard, can hardly be overemphasized" (31). The narrative even pauses for a verse—while Tarquin blinks—to mark the significance of the impending specular violation. Had Tarquin's eyes remained shut,

> Then Collatine again by Lucrece' side
> In his clear bed might have reposed still.
> But they must ope, this blessed league to kill;
> And holy-thoughted Lucrece to their sight
> Must sell her joy, her life, her world's delight. (381–85)

There is a prurient, breathless quality to the verse as it pores over her body ("Her azure veins, her alabaster skin, / Her coral lips, her snow-white dimpled chin" [419–20]). In effect the reader becomes, with Tarquin, a voyeur, watching Lucrece with "lewd unhallowed eyes" (392).[18]

Significantly, however, the description of Lucrece's posture is inconsistent. Her breasts—"like ivory globes circled with blue" (407)—are emphatically subject to Tarquin's gaze, and she wakes when she feels Tarquin's hand upon "her bare breast" (439). His hand remains there (463) as his tongue begins "to sound a parley" to Lucrece, who, we learn, "o'er the white sheet peers her whiter chin" (471–2). This physical impossibility—we have to imagine the breast as at once erotically exposed and modestly covered—points to Shakespeare's ambivalent presentation of his heroine: She is both sexually provocative and sexually chaste—a paradox epitomized in the poet's apostrophe to her hair that "like golden threads play'd with her breath": "O modest wantons, wanton modesty!" (400–1).

The ambivalence that marks Shakespeare's description of Lucrece's body characterizes his presentation of her throughout the poem. On one level he asserts Lucrece's innocence (1254 ff.), and supports this with frequent religious imagery. Lucrece is an "earthly saint adored by this devil" (85), "holy-thoughted" (384) and "divine" (193). Like the lamb and dove with which she is associated (167, 360, 737), she is a sacrificial victim ("My blood shall wash the slander of mine ill; / My life's foul deed my life's fair end shall free it" [1207–8]). On another level, however, Shakespeare implicitly confirms Tarquin's rationale of sexual provocation. "[T]he fault is thine," Tarquin declares (482), "Thy beauty hath ensnar'd thee [. . .]" (485);[19] and we are prepared to accept this claim by the narrator's association of beauty and guilt from the first entrance of Lucrece, "[w]ithin whose face beauty and virtue strived" (52). The strife of beauty and virtue, which Shakespeare develops in the following stanzas, is more than a conceit. It implies a radical conflict. Beauty is linked with the red of Lucrece's blushes (blood/heat/passion), virtue with the white of her skin (cold/purity). The imagery of red and white, rose and lily pervades the poem. When Lucrece asks Tarquin "[u]nder what colour he commits this ill" (476), he replies:

> The colour in thy face,
> That even for anger makes the lily pale
> And the red rose blush at her own disgrace,
> Shall plead for me and tell my loving tale? (477–80)

After the rape, Lucrece's blushes before her groom mark her conscious guilt, telling the tale of the rape (1352–55). Ultimately, the blush of guilty beauty is transformed into the "crimson blood" (1738) with which she atones for that guilt and for Tarquin's "scarlet lust" (1650).

Shakespeare represents the rape as a moral exchange: Lucrece loses her honor—that is, *Collatine's* honor (834)—figured repeatedly as a jewel or treasure (34, 1056, 1191), and absorbs Tarquin's guilt ("She bears the load of lust he left behind" [734]). Tainted by this lust, her body becomes a "blemish'd fort" (1175), a "poison'd closet" (1659), a "polluted prison" (1726), and an object of her shame and hatred. She tears her flesh with her nails (739), beats her breast (759), and fears the exposure of the coming day (746–56). Her ambiguous position—innocent soul in a guilty body—is epitomized in the transformation of her blood after death: Part remains "pure," part is "stain'd" (1742–43).[20]

Shakespeare seems to have been particularly interested in the response of his heroine to her rape. In contrast to Lavinia, who has no chance to voice her grief, Lucrece has over 600 lines—almost a third of the poem. In addition to her sense of shame and guilt, two aspects of the ravished Lucrece are striking. The first is the violence of her anger: She curses Tarquin fiercely (967–1001),[21] tears the image of Sinon—for her a Tarquin figure—with her nails (1562–66), and makes the lords swear an oath of revenge (1688–94) before instructing them by her suicide. This Lucrece is dangerous as well as pathetic.

Closely related to Lucrece's anger is her hysteria. Brooding over her "cureless crime" (772), Lucrece is consumed with grief:

> No object but her passion's strength renews,
> And as one shifts another straight ensues.
> Sometime her grief is dumb and hath no words,
> Sometimes 'tis mad and too much talk affords. (1103–6)

Her response to the picture of Troy continues the representation of her dementia ("Show me the strumpet that began this stir, / That with my nails her beauty I may tear!" [1471–72]). The "joyless smile" with which Lucrece rejects the lords' verdict of "innocent" and prepares to kill herself is perhaps the most chilling instance of her madness (1711). Her psychological imbalance is the mental equivalent of Lavinia's mutilation: They are both represented as in some way monstrous.[22]

Shakespeare moves deftly from the private dimension of the crime to its political consequences as Brutus plucks the knife from Lucrece's body (1807). His rebuke to the wailing Collatine extends to Lucrece:

> Is it revenge to give thyself a blow
> For his foul act by whom thy fair wife bleeds?
> Such childish humour from weak minds proceeds;
> Thy wretched wife mistook the matter so,
> To slay herself that should have slain her foe. (1823–27)

Brutus's trenchant criticism forcefully juxtaposes the masculine world of public activity with Lucrece's domestic confines. Given her gender, the most Lucrece can do is to transcend it in heroic suicide: Slaying the foe is clearly not an appropriate option. Given Brutus's gender, he can and does take up arms against the foe. The last verse solemnly records the outcome: "The Romans plausibly did give consent / To Tarquin's everlasting banishment" (1854–55). Since Lucrece had called for Tarquin's blood (1686), this is both less and more than she asked for. Instead of private revenge the men effect a change to the political constitution. The discrepancy points to the political appropriation of a woman's violation by men that is a consistent feature of the Lucrece story and its analogues.

The Rape of Lucrece

In contrast to Shakespeare's poem, Heywood's play presents a broadly popular dramatization of Roman history.[23] Its focus is not the characterization of the heroine—characterization is indeed minimal—but a larger vision of a commonwealth's sickness and recovery. We see the progressive deterioration of a community under a tyrant—a process that reaches its nadir in the rape—and the reversal that follows it.

Within Heywood's vision of the community, women occupy clearly defined roles. Tullia, the old king's daughter (and mother of Sextus Tarquin, the rapist), is an evil usurper of masculine authority, intruding into the world of public affairs. Lucrece is her opposite: Virtuous and passive, she remains in her proper domestic sphere. This pairing of females—predator and victim—recalls Shakespeare's use of Tamora and Lavinia, Lady Macbeth and Lady Macduff. In each case a community's evil begins with one female's aggressive and ambitious interference in the public realm and leads to the suffering of the second innocent woman: As with Eve and Mary, the original sin of the first is redeemed through the second. The innocents in these plays, however, atone with death rather than with birth.

Heywood found the hint of this structure in his source. Livy attributes the fall of the Tarquins to the tragic guilt incurred by Tullia (bk. 1, ch. 46). In Livy, as in the play, she is the root of all evil, goading her husband into rebellion. Heywood, however, heightens her role considerably, bringing her into dramatic prominence from the opening scene. Repeated imagery of childbirth underscores the link between her sexuality and her ambition: "With ardency my hot appetite's a fire," she declares, "Till my swolne fervor be delivered / Of that great title Queene" (166). Inordinate sexual desire is here a figure for dangerous political desires. The point is emphasized by Collatine's prophetic remarks, just before the deposition: "[S]ome strange project lives / This day in Cradle that's but newly borne" (167). "The wife of *Tarquin* would be a Queen," he continues, "nay on my life she is with childe ill she be so" (168). Valerius concurs: Tullia "longs to be brought to bed of a Kingdom" (168). This misogynist discourse constructs female sexuality itself—in its agency, its desire, its procreative powers—as something fearful, potentially demonic. Figuratively, then, Tullia's corrupt sexuality breeds a diseased body politic: As Valerius observes, the "commonwealth is sick of an Ague, of which nothing can cure her but some violent and sudden affrightment" (168).

Tullia's sin of political ambition is compounded by parricide. Unlike Livy, Heywood gives Tullia a leading part in the deposition scene. She contemptuously defies her father and, rejoicing in his death ("We have our longing"[172]), literally treads upon his body. Heywood here elicits the same horror as Shakespeare does through Goneril and Regan's abuse of Lear—the horror of a subject daughter's "unnatural" rebellion against a father king. Brutus's response, like Lear's, emphasizes the relation between gender and crime:

> but the Queene,
> A woman, fie fie: did not this shee-paracide

> Adde to her fathers wounds? and when his body
> Lay all besmeard and staynd in the blood royall,
> Did not this Monster, this infernall hag,
> Make her unwilling Chariotter drive on,
> And with his shod wheeles crush her Fathers bones?
> Break his craz'd scull, and dash his sparckled braines,
> Upon the pavements, whilst she held the raines!
> The affrighted Sun at this abhorred object,
> Put on a maske of bloud, and yet she blusht not. (174)

Tullia's refusal to blush signifies her utter abandonment of feminine propriety (a modest blush is the sign of shamefastness, a woman's acknowledgment of her guilty and inferior sexuality), and her criminal movement from domestic confines to public conflict.

Though Lucrece herself does not appear for some time, the contrast between the two women is made explicit immediately after the deposition (172), and numerous references, fraught with irony, keep her in our minds. The scenes in which Lucrece does appear before the rape (again, Heywood's invention) illustrate her domestic virtues as prudent housewife and decorous spouse. Indeed, she is almost ludicrously exemplary, speaking in aphorisms: "I am opinion'd thus: Wives should not stray / Out of their doors their husbands being away" (210); "With no unkindnesse we should our Lords upbraid, / Husbands and Kings must always be obaid" (214). While it is difficult for a modern reader to take this Lucrece seriously, nothing in the context suggests that Heywood's representation of her is deliberately parodic.[24] In her distress Lucrece is more interesting: She moves beyond aphorisms to echo Shakespeare's Lucrece. Like her predecessor, Heywood's Lucrece takes the guilt of the rape upon herself. It is her "blot," her "scandall," her "shame" (235). In taking her chastity, Tarquin has robbed her of herself—her identity as "soule-chast Lucrece" (224)—and made her monstrous: "Being no more a woman, I am now / Devote to death and an inhabitant / Of th'other world" (235).

Lucrece's function as sacrificial victim is foreshadowed early in the play. Her rape is the "violent and sudden affrightment" that Valerius diagnoses as the only remedy for the commonwealth's "Ague" (168). The Oracle at Delphi declares, "Then *Rome* her ancient honours wins, / When she is purg'd from *Tullia's* sins" (184); and as the play continues, there are frequent suggestions of a providential order asserting itself, moving towards that purgation. The notion of divine ordinance is strengthened by Lucrece's poignant prayers before and during the rape (once more, Heywood's invention). Going to bed, she says, "my hearts all sadnesse, / *Iove* unto thy protection I commit / My chastitie and honour to thy keepe" (220). Just before Tarquin carries her off, Lucrece cries, "*Iove* guard my innocence"; to which her ravisher replies, "*Lucrece* th'art mine: / In spight of *Iove* and all the powers divine" (225). Heywood is indicating not divine indifference but divine will. Lucrece's pathetic complaints after the rape only emphasize this:

> oh you powerfull Gods,
> That should have Angels guardents on your throne.
> To protect innocence and chastitie! oh why
> Suffer you such inhumane massacre
> On harmlesse vertue? wherefore take you charge,
> On sinlesse soules to see them wounded thus
> With Rape or violence? or give white innocence,
> Armor of proofe gainst sinne: or by oppression
> Kill vertue quite, and guerdon base transgression?
> Is it my fate above all other women? (234)

In the context of the play, the answer to the last question must be "yes." It is the fate of Lucrece, alone of all her sex, to redeem Rome with her innocent blood.

In spite of Heywood's large debt to Shakespeare's poem,[25] his staging of the rape is as innocent of eroticism as Shirley's treatment of sexual violence in *The Martyred Soldier.* His Tarquin does pause to study Lucrece before waking her, but we hear no sensual description of her body. His deliberation is perfunctory, his appraisal of Lucrece the most general:

> Oh who but *Sextus* could commit such waste?
> On one so faire, so kinde, so truly chaste?
> Or like a ravisher thus rudely stand,
> To offend this face, this brow, this lip, this hand? (222)

The dialogue that precedes the rape seems pitched to arouse pity and horror rather than sexual excitement. Lucrece weeps and pleads ineffectually before the obdurate Tarquin carries her out.

Nevertheless, there is an ambivalence in the play's treatment of sexuality and of the rape itself. This is manifest in part through Heywood's use of song. Of the nineteen songs included in the text, seven comment either directly or obliquely on the political theme.[26] A further four are merely ballads—their inclusion illustrates the idleness of the Roman lords, who have abandoned public life because of Tarquin's tyranny.[27] One song—*"Packe cloudes away, and welcome day"* (227)—Heywood, with superb irony, places immediately after the rape.[28] However, seven of them are bawdy. While it is true that their inclusion can also be explained as symptomatic of the court's moral dissolution, their presence is jarring. They cater to the appetite the play ostensibly decries.

This is particularly true of the song which Horatius, Valerius and the Clown sing together soon after the rape. Heywood takes a great deal of trouble to set this song up. When the Clown, as Lucrece's messenger, brings her summons to the camp, Brutus, Collatinus and Scaevola rush off immediately (231). Horatius and Valerius, however, remain on stage to press the Clown for information. Clearly they have guessed what has happened and foresee the result. Horatius cries:

> The newes, the newes, if it have any shape
> Of sadnesse, if some prodegie have chanst,
> That may beget revenge, Ile cease to chafe,
> Vex, martyr, grieve, torture, torment my selfe,
> And tune my humor to strange straines of mirth,
> My soule divines some happinesse, speake, speake:
> I know thou hast some newes that will create me
> Merrie and musicall for I would laugh,
> Be new transhapt [. . .]. (231)

In other words, the rape is just what the Roman lords have been waiting for, the "prodegie" that will "beget revenge." Significantly, Horatius feels he can now stop tormenting himself: By implication he will direct his anger to its proper cause, the Tarquins. As we know, however, Lucrece will "[v]ex, martyr, grieve, torture" and "torment" herself instead. When the Clown declares that he has sworn to say nothing of what has happened, Valerius urges him to sing the news, and invites Horatius to second his appeal (232). By this clumsy device Heywood sets up the three part "catch," *"Did he take faire Lucrece by the toe man?"* (232–33). The song titillates the audience by successively naming the parts of Lucrece's body from toe to thigh. In response to each sung question, the Clown laughs an assent. Thus it begins:

> VAL. Did he take faire Lucrece by the toe man?
> HOR. Toe man.
> VAL. I man.
> CLOW. Ha ha ha ha ha man.
> HOR. And further did he strive to go man?
> CLOW. Goe man.
> HOR. I man.
> CLOW. Ha ha ha ha man, fa derry derry downe ha fa derry dino. (232)

Presumably the stage business would include bawdy gestures in accompaniment. In effect, the song reenacts the rape in a jocular context.

As Donaldson observes, "jokes about rape have a special quality: they characteristically imply that the crime may not in fact exist; that it is a legal and social fiction, which will dissolve before the gaze of humor and the universal sexual appetite" (89). I think this is the effect of Heywood's song. We are invited to laugh about the rape: It becomes, temporarily, a comic seduction, a humorous exploit. It is possible to read this "catch," with the other bawdy songs, as exemplifying the lust of which Rome must be purged. This seems to me improbable, however. As the title page of the 1638 edition make clear, the songs were one of the play's major attractions.[29] Whatever Heywood's intentions, it seems unlikely that his audience responded critically to the songs. On the contrary, there was apparently a great demand for the lyrics. I suggest then that the inclusion of the bawdy songs was a shrewd and successful theatrical move; that the tension between these songs and the overt theme of the play reveals an ambivalence about rape characteristic not just of Heywood's play, but of social attitudes in a patriarchal society (rape matters very much/rape does not matter at all); and finally, that for the common Jacobean auditor at the Red Bull, there would be no tension. The same audience could respond with pity to Lucrece's distress in one scene and with complicit laughter to the bawdy "catch" in the next. Heywood has it both ways.[30]

Certainly the rape recedes in importance before the renewed heroism of the Romans. In Livy this renewal springs directly from Lucrece's death. It is signified by Brutus's sudden transformation from fool to hero as he pulls the knife from Lucrece's body and makes the witnesses swear the oath of rebellion. In Heywood's play this call-to-arms precedes Lucrece's suicide and Brutus uses his own sword. Heywood thus deprives Lucrece of her traditional glory: Her death does not empower the Romans in the same way. In Livy Lucrece's "masculine" heroism instructs the Romans, her dripping knife is a symbol of their remasculinization. In Heywood Brutus upstages Lucrece; *his* heroism owes nothing to *hers, his* sword symbolizes the Romans' renewed virility. Lucrece's suicide is almost irrelevant. Indeed, she kills herself while *"The Lords whisper"* (238), presumably about the rebellion. Their acknowledgment is the most cursory (roughly 17 words). Brutus just says, bluntly, "She's dead," before warming to his political oratory ("then turn your funerall teares to fire / And indignation [. . .] " [238]). The scene ends with jubilant preparations for war: The directions call for *"A great shout and a flourish with drums and Trumpets"* (240), before Brutus delivers his stirring exit line: *"Iove* give our fortunes speed. / Weele murder, murder, and base rape shall bleed" (240).

According to Livy, Sextus Tarquin, the rapist, met an ignominious death: He fled to a nearby town where he was assassinated by his former allies (bk. 1, ch. 60). In the year following his death, and the expulsion of his family from Rome, the young republic fought a series of battles to maintain its freedom. The famous episodes of Horatius at the bridge and Scaevola before Lars Porsenna belong to this period. Heywood incorporates these two highlights of Roman heroism into his action-packed fifth act and keeps Sextus alive and valiant for a final climactic fight. For like the Romans, Sextus too becomes heroic. In the last battle, he breathes defiance to all and, invoking the example of Hector and Achilles, challenges Brutus to single combat. Brutus accepts in terms that indicate a reciprocal recognition of heroic stature: "Hadst thou not done a deed so execrable / That gods and men abhorre, ide love thee *Sextus,* / And hug thee for this challenge breath'd so freely" (251). Sextus's reply denigrates Lucrece even as it exalts Brutus:

> To ravish *Lucrece,* cuckold *Collatine,*
> And spill the chastest blood that ever ran
> In any Matrons vaines, repents me not
> So much as to ha wrong'd a gentleman
> So noble as the Consull in this strife. (252)

Clearly Sextus is speaking as a villain whose values are vitiated by his evil nature. Nevertheless, this male bonding, consummated by the combatants' mutual destruction—the directions make it clear they die together[31]— inevitably modifies our experience of the play. It is the culminating event and has all the force of its closing position. It is with this glorification of male heroism, rather than the vindication of female integrity, that the weight of the play finally lies.

Appius and Virginia

Appius and Virginia, as Lucas remarks, "seems archaic for its date."[32] Even if that date is 10 or 15 years earlier than Lucas believed, his observation holds true. Like R. B.'s *Apius and Virginia,* Webster's drama has the simplicity of a morality play. In its treatment of gender, however, it is close to *The Rape of Lucrece.* Here, as in Heywood's play, the heroine's role is overshadowed by that of the avenging males. Virginius, not Virginia, is the central figure in this drama.

Here, as in R. B.'s play, the lines between good and evil are clearly drawn, and characters have an emblematic value: Appius is a figure of Lechery, while his assistant, Clodius, acts as pander and Vice, pulling the plot strings and providing black comedy. Virginia, as her name implies, is Chastity, her father Honesty. This moral structure is reflected in the play's language: Whenever the central issue of Appius's designs arises, the good characters burst into invective. According to Icilius, Appius is a "Lustful Lord" (2.3.90), "a divel, a plain divel" (3.1.58) and an "incontinent, loose Leacher" (3.1.61). "All the fire in hell," he declares, "Is leapt into his bosom" (4.1.42–43). Virginia swears "by a Virgins honour and true birth," "I never had a dream / So terrible as is this monstrous divel," Clodius (3.2.297–99). As the confrontation with Appius approaches, Rome itself seems to dissolve into the underworld, with the tyrant, like Rhadament, sitting in judgement over it (4.1.9–12). "Where are we," Icilius asks, bewildered, "—in a mist?—or is this hell?" (3.2.399). In the end, however, after the pattern of the morality play, the corrupt judge is justly doomed. Appius himself dutifully draws the moral of the story before he accepts his death (5.2.140–44).

Like R. B., Webster enhances the hagiographic elements in the original story by enlarging Virginia's role.[33] He pointedly illustrates Virginia's free choice of her honorable suitor, Icilius (1.4.128–32), and her spirited rejection of Appius. When Clodius appears at Virginia's side to woo her with music and promises of power, wealth and pleasure, Virginia rebukes her tempter with asperity: "Let thy Lord know, thou Advocate of lust, / All the intentions of that youth [Icilius] are honourable, / Whil'st his are fill'd with sensuality" (2.1.79–81). (Apart from this kind of righteous indignation Virginia says little: On the whole she is silent as well as chaste and obedient.) Her last words are a prophetic indictment of Appius, and, like those of a martyr, bear witness to her faith:

> Remember yet the Gods, O *Appius,*
> Who have no part in this. Thy violent Lust
> Shall like the biting of the invenom'd Aspick,
> Steal thee to hell. (4.1.256–59)

Most importantly, Virginia asks for death, deliberately choosing her martyrdom:

> O my dear Lord and father, once you gave me
> A noble freedom, do not see it lost
> Without a forfeit; take the life you gave me
> And sacrifice it rather to the gods
> Then to a villains Lust. (4.1.31–35)

Virginia's function as a sacrificial victim, like that of Heywood's Lucrece, is foreshadowed by imagery of the state's disease, "incurable, / Save with a sack or slaughter" (1.4.82–83). After her death, the spectacle of her father's appearance keeps the sacrifice visually prominent. He enters the camp, *"with his knife, that and his arms stript up to the elbowes all bloudy"* (4.2.85.s.d.). During his oration to the soldiers Virginius explicitly draws attention to his bloody limbs (126–27, 130–32)—an attention inevitably reinforced by every gesture—and concludes by claiming it his pride to have "bred a daughter whose chast blood / Was spilt" for Rome (206–7). In the final scene the spectacle of blood culminates in the display of Virginia's corpse. Icilius brings it on to "anatomize" the sin of Appius (5.2.79), and it bleeds again in the villain's presence. Finally, Virginia's body, an icon of her martyrdom, is solemnly borne from the stage. Virginius, who has the last word, makes the connection with Lucrece: "Two [Ladies fair, but] most infortunate, / Have in their ruins rais'd declining *Rome*—/ *Lucretia* and *Virginia,* both renown'd / For chastity" (5.2.192–95).

In spite of this tribute, and in spite of the significantly larger part Webster assigns her, Virginia has only a minor role in the play. As in *Titus Andronicus* the focus is not on the daughter's suffering, but on the father's. It is the pathos of *his* grief that Webster exploits in the death scene. Indeed, after Virginia's prophetic rebuke to Appius (quoted above), she is silent for almost 200 lines— while the men fight over her—before dying without a word. It is Virginius who articulates his sorrow, and he does so at considerable length: "Farewel my sweet *Virginia,* never, never / Shall I taste fruit of the most blessed hope / I had in thee" (4.1.321–23), and so on, for several more maudlin lines (324–36). Just as Lavinia's suffering is subordinate to her father's, so here Virginia's pain is subsumed in that of Virginius. This is all the more remarkable since Virginia, unlike the ravished Lavinia, has a tongue, and Virginius—unlike Titus—is the principal agent of his daughter's suffering.

This emphasis on Virginius's grief is one of the ways in which Webster elicits sympathy for his hero and forestalls censure of the parricide. In the sources there is no suggestion of blame for Virginius.[34] The witnesses, as Painter translates it, "cryed out upon the wickednesse of Appius" and "deplored the necessitie of the father" (1: 39). Webster, however, could anticipate more resistance from *his* audience.[35] In addition, then, to a saintly Virginia who actively begs for death, he presents a Virginius tortured by his responsibility for that death—a martyr rather than a murderer. After the slaying Virginius suffers a mental and physical deterioration; he *"Offers to kill himself"* (4.2.188.s.d.) before the army; he is wracked by a fever and seems to be haunted by a spirit (5.1.76–78, 109). There is suggestion of Lear about him at this point—the noble soul worn out with grief. When Icilius tries to comfort him ("There's hope yet you may live / To outwear this sorrow" [177–78]), Virginius replies: "O impossible. / A minutes joy to me, would quite crosse nature" (179–80).

Webster also attempts to incorporate and dispel any condemnation of his hero by including what is, in effect, Virginius's trial by the jury of the army. When he arrives in the camp, covered in blood, Virginius begins to tell commanding officer the story of the slaying. Minutius's initial response ("Most wretched villain!" [133]) provides a release for the audience's horror. The emotional balance shifts, however, as Virginius pleads his case eloquently:

> Alas, might I have kept her chaste and free,
> This life so oft ingag'd for ingrateful *Rome,*
> Lay in her bosom. But when I saw her pull'd
> By *Appius'* Lictors to be claim'd a slave,
> And drag'd unto a publick Sessions-house,
> Divorc'd from her fore-Spousals with *Icilius,*
> A noble youth, and made a bond-woman,
> Inforc'd by violence from her fathers armes
> To be a Prostitute and [Paramour]
> To the rude twinings of a leacherous Judge;
> Then, then, O loving Souldiers, (I'l not deny it)
> For 'twas mine honor, my paternal pity,
> And the sole act, for which I love my life . . .
> Then lustful *Appius,* he that swayes the Land,
> Slew poor *Virginia* by this fathers hand. (142–56)

The common soldiers' response cues our own: "O villain *Appius,*" they cry, "O noble *Virginius*" (157–58). Virginius then appeals to the men for a judgement: "Did *Appius* right, or poor *Virginius* wrong? / Sentence my Fact with a free general tongue" (160–61). Unhesitatingly they pronounce Appius "the Parricide" and "*Virginius* guiltless of his daughters death" (162–63). This trial by the jury of common soldiers contrasts with and corrects the earlier trial by the corrupt judge Appius. On the balance, then, Webster presents the killing as justifiable parricide—terrible, but necessary. The guilt of the deed attaches to Appius, who "Slew poor Virginia" by her "fathers hand." Virginius is a heroic figure, made tragic by the "necessitie" that compels him to kill his daughter, supported by his fellows and finally vindicated by the success of his rebellion.

As in Heywood's play, the heroine's death recedes in importance before the male bonding that it provokes. Indeed, the really crucial relationship in this play is not the one between father and daughter—though obviously that is important—but the homosocial bond between Virginius and his soldiers. One of Webster's additions to the story is a famine in the army: The men have not been fed properly in months. This crisis allows Virginius to emerge as the champion of the common soldier. We witness his eloquent appeal to the senate for supplies (1.4.6–33), and—when Appius has haughtily refused them—Virginius's resolution to beggar himself to feed his men (1.1.117–20).

The intensely emotional relationship between the captain and his soldiers resembles that of a patriarchal family. Indeed the general, Minutius, in a brief homily on obedience, draws the analogy between the hierarchy of the commonwealth and the hierarchy of the army: "[E]very Captain / Beares in his private government that forme, / Which Kings should ore their Subjects, and to them / Should be the like obedien[ce]" (2.2.185–88). Virginius, as Webster makes clear, is the ideal patriarch: While he maintains a wholesome discipline, he is wholly devoted to the care of his men.

There are two big scenes in the camp: They balance each other, one on either side of Virginia's death. The second scene, described above, Webster inherited from his sources; the first he invented. In the earlier scene Virginius returns from his unsuccessful trip to Rome to find his men in a mutiny (2.2.96.s.d.). At this point, before the provocation of Appius's attempt on Virginia, Virginius refuses to countenance rebellion. In an elaborate display of his authority, he first quells the insurrection—his mere presence is enough—demonstrates to the watching general his soldiers' renewed obedience, then angrily discharges them all (113–47). The soldiers beg his forgiveness: "Dear Captaine!" (159), they protest, "wee'l starve first, / Wee'le hang first, by the gods, doe any thing, / Ere wee'le forsake you" (163–65). When at length Virginius permits the general to reconcile him with his men, Minutius does so with a formal declaration: "We therefore pardon you, and doe restore / Your Captaine to you, you unto your Captaine" (211–12).

This early divorce and reconciliation between Virginius and his soldiers foreshadows the events that follow Virginia's death, when the hero is first estranged from the men, then vindicated by them. "Vindicated" is indeed too mild a word: In their adulation they proclaim him general (4.2.178) and hail him as their destined preserver (203). This homage, which concludes the fourth act, is arguably the emotional climax of the play.

Just as the soldiers' "free general tongue" exonerates Virginius of guilt in his daughter's death, so too it relieves him of any taint of ambition in the rebellion. He is persuaded only with difficulty to assume command of the insurrection: It requires the united efforts of the soldiers and Minutius to prevail upon him (4.2.177–203). When Virginius finally accepts their choice, he does so in the hope of protecting them all from his "inhumane sufferings" (204–5). His willingness to lead them thus springs from the purest of motives, his devotion to his men. Though Virginius gestures at suicide (4.2.187–88), he lives to prosecute his revenge reluctantly, a dutiful patriarch.

For a modern reader the defense of male property that underlies the ideology of chastity emerges starkly in Webster's play. Despite religious accretions, the story provides a paradigm for the ancient legal definition of rape—that is, theft of a woman. Appius's legal maneuver is a sophisticated form of abduction, a violation of the socially ratified means of acquiring a woman. In effect, Appius is attempting to cheat Virginius of his bride-money and Icilius of his purchase. Had he succeeded, Virginia's market value would have dropped sharply. Thus, after the killing, Virginius is able to silence Icilius's reproaches by retorting: "Had my poor girle been ravish'd, [. . .] Your love and pity quickly had ta'ne end" (5.1.130–32). It is as property owners that Virginius addresses the army:

> You that have wives lodg'd in yon prison *Rome,*
> Have Lands unrifled, houses yet unseis'd,
> Your freeborn daughters yet unstrumpeted,
> Prevent these mischiefs yet while you have time. (4.2.198–201)

The hero's proprietorial stake in the heroine's death, effectively cloaked in *The Rape of Lucrece,* becomes transparent here.

As I claimed in my introduction, these plays, based closely on what I have called the "classical paradigm" of rape, do not, as Gossett suggests, "imply a concern and respect for women in general" (326); nor, in spite of the serious consequences that follow rape or attempted rape, do they imply "that women have a personal integrity that cannot survive violation" (324). Rather, these classically derived plays demonstrate that women have a value as male property that cannot survive violation. Moreover, in dramatizing the classical paradigm of sexual assault—a model dignified by its antiquity and deeply appealing in its melodrama—they represent the attack as socially functional. The assaulted woman embodies a community in thrall to a tyrant, and the sexual depravity of the rulers—aggressive, lustful women and effeminate, lustful men—symbolizes the evils of the government. In contrast to the "saint's plays," in which the triumphant chastity of the heroine symbolizes a spiritual strength, these plays emphasize the necessity of the heroine's death and the rebellion that ensues. Female martyrdom leads to a remasculinization of the community, demonstrated in the valor of the insurgent males. In *Appius and Virginia,* as in *Titus Andronicus,* the suffering of the heroine is ancillary to the pain of her father: He becomes an object of tragic pity primarily through his grief for her. In the plays by Heywood and Webster the homosocial bonding that follows the heroine's death is particularly intense—even, at moments, *exuberant.*

Virginia dies chaste, but Lavinia and Lucrece (both Shakespeare's and Heywood's) reveal—as the heroines of the "saint's plays" cannot—the consequences of the loss of physical chastity for the heroine. In each case she becomes, like the ravished Medusa and Philomela, monstrous in some way. In patriarchal terms a raped woman is literally monstrous—that is, *unnatural* (*OED* 1a), a deviation from the natural order of things—and this deviation manifests itself in aberrant behavior and/or appearance. Lavinia—grotesque and frightening as she pursues Marcus or holds the bowl to catch her assailants' blood—prefigures the various sinister metamorphoses of the raped woman on the Jacobean stage.

Chapter 4 ⊛

"Some Injury in a Matter of Women": Variations on the Classical Theme

[. . .] it appears that women have caused much destruction, have done great harm to those who govern cities, and have occasioned many divisions in them; for, as we see in this *History* of ours [Livy], the outrage to Lucrece took their position from the Tarquins. That other outrage, to Virginia, deprived the Ten of their authority. So Aristotle gives among the first causes for the falls of tyrants some injury in a matter of women, either by whoring them, or raping them, or breaking off marriages [. . .]

—Machiavelli, *Discourses* ("How a State Falls Because of Women")

Machiavelli's remarks on women and civil unrest point both to the misogynist tendency to blame the victim of sexual assault—it is women who "have caused much destruction," not the men who committed "the outrages"—and to the persistence of the classical paradigm of rape in the Renaissance. Congruent with Machiavelli's political analysis, early modern drama most frequently represents sexual violence in terms of its effect on male bonds: In Heywood and Webster's direct adaptations of Livy, examined in the previous chapter, the disruption of the patriarchal political order "caused" by the sexually threatened women is ultimately beneficial to the community. Similarly, in *The Revenger's Tragedy* (1606) and *The Second Maiden's Tragedy* (1611), where the classical paradigm is united with hagiography and an Italianate revenge plot, the political consequences of sexual assault help to regenerate a corrupt state. In *Cymbeline* (1609), too, where Shakespeare recalls his poetic treatment of the Lucrece myth, Imogen's sexual victimization promotes the renewed health of Britain. However, in three later plays—Fletcher's *Valentinian* (c. 1612), Rowley's *All's Lost by Lust* (1619–20), and Middleton's *Hengist, King of Kent* (1619–20)—the classical pattern of female sacrifice and male valor is subject to increasingly ironic modifications. In *Bonduca* (c. 1613) Fletcher subverts the paradigm in ways that illuminate all the "Lucrece" plays, clearly demonizing the raped women who "cause much destruction" to their community and celebrating the male homosocial bonds that survive.

The Revenger's Tragedy

In *The Revenger's Tragedy*[1] the Lucrece/Virginia myth is a strong but oblique presence: The heroine is neither ravished nor slaughtered, and the machinations of the revengers displace the martial heroism that Heywood and Webster exploit so fully. Nevertheless, within the realm of Italianate intrigue the classical paradigm serves as a powerful touchstone for female virtue and the political exigencies of tyranny.

The playwright complicates the "Lucrece" paradigm by multiplying the tyrant and the crime: There are three (male) sexual predators, three assaults, and three saintly victims in various stages of martyrdom. Gloriana's skull—exhibit A for Vindice's prosecution—is a vivid memento of her murder by the Duke nine years ago for refusing his sexual demands (2.3.130–31). In the immediate past, the Duchess's youngest son has "played a rape on Lord Antonio's wife" (1.1.109), and she, like Lucrece, has killed herself.[2] Castiza, pursued by the Duke's son Lussurio, is a candidate for future martyrdom.

Like Gloriana, Antonio's wife died of poison, and the display of her body—exhibit B—provides the climax of the first act. Antonio invites the lords to behold the spectacle and highlights significant details:

> A prayer book the pillow to her cheek;
> This was her rich confection, and another
> Placed in her right hand with a leaf tucked up,
> Pointing to these words:
> *Melius virtute mori, quam per deducus vivere.*[3] (1.4.14–18)

The scene is an emblem of chastity, a classical Lucrece made Christian with a prayer book (Robertson 221). While we gaze at the tableau, Antonio tells the story of the rape. It is considerably more horrific than Tarquin's crime because it is blatant. While Tarquin, craving secrecy, used stealth and darkness, the Duchess's son assaulted his victim in public, in the "artificial noon" of torchlight (1.4.38–45). However, the resemblance to the Lucrece story is underlined by the lords' admiration for her death ("A wondrous lady of rare fire compact, / She's made her name an empress by that act" [49–50]), and their consequent oath of revenge sworn on a drawn sword (57–65). The whole scene, from the discovery of the body to the oath that concludes it, telescopes the story of Lucrece into 78 lines. Although technically this oath—in contrast to that of Brutus—leads nowhere,[4] it substantiates Hippolito's claim of widespread support ("five hundred gentlemen" [5.2.26]) for their coup in act 5.

Lussurio's pursuit of Castiza is the third sexual crime to prompt an overdetermined rebellion. Significantly, she is not at court, and her physical distance from the palace, the seat of corruption, indicates a moral distance. The playwright injects novelty into the conventional temptation by having her brother, Vindice, act as pander. (Lussurio has employed him, as the malcontent "Piato," to procure Castiza. Failing that, he is on oath to bribe her mother, Gratiana, to their aid.) Although Vindice has qualms about his unnatural commission, he embraces the opportunity to test the virtue of his mother and sister (1.3.176–186). (Significantly, neither his brother nor his mother, when she learns the truth, challenges Vindice's right to subject the women to this test.)

The temptation plot forms a self-contained morality play in two acts. In the first (2.1), the playwright displays Castiza's extraordinary chastity, Vindice's moral ambivalence, and Gratiana's common venality. When Vindice appears in disguise to give Castiza a letter from the Duke, she responds—like Virginia to Clodius—in the tradition of the virgin martyrs, with righteous indignation. She even strikes him (30.s.d.). Her verbal reply underlines her singularity: "Tell him my honour shall have a rich name / When several harlots shall share his with shame" (37–38). Both Vindice and his mother reflect upon the action in misogynistic terms: We are prompted by both to see Gratiana's fall from grace in terms of her gender. Gratiana remarks, "Men know, that know us, / We are so weak their words can overthrow us" (105–6); and Vindice declares in disgust,

> Were't not for gold and women there would be no damnation,
> Hell would look like a lord's great kitchen without fire in't;
> But 'twas decreed before the world began
> That they should be the hooks to catch at man. (254–57)

Gratiana, we understand, is typical of her sex; Castiza is the exception that proves the rule. We are, moreover, implicitly invited to contrast the moral strength of Castiza's virginity with the weakness of her widowed mother's sexuality.

In the second "act" of this morality play (4.4), after Gratiana repents in tears (38–39), Castiza announces the moral of the story in terms that implicitly respond to her mother's earlier observation on the corrupting power of eloquence:

> If maidens would, men's words could have no power;
> A virgin honour is a crystal tower
> Which being weak is guarded with good spirits:
> Until she basely yields no ill inherits. (151–154)

This is the sexual morality of the saints' lives: Since God protects virginity, the woman who does succumb to sexual coercion—who "basely yields"—is responsible for her own fall. Gratiana's delighted response emphasizes, once more, the singularity of Castiza's virtue: "Oh happy child! Faith and thy birth hath saved me. / 'Mongst thousand daughters happiest of all others! / Be thou a glass for maids, and I for mothers" (155–57).

The misogyny latent in this idealization of feminine virtue is manifest in the playwright's use of Gloriana's skull. Logically the skull, like the body of Lord Antonio's wife, should be a symbol of moral purity, since Gloriana died through her virtuous resistance to the Duke's lust. The dramatist uses it, however, as an image of corruption, a sign not only of the Duke's sins, but—in the largest sense—of all sin and the physical and spiritual death attendant upon it. More specifically it signifies the deadly snare of female flesh. This transference of guilt from the Duke to his victim begins even in the play's opening passage, as Vindice recalls Gloriana's face:

> [. . .] 'twas a face
> So far beyond the artificial shine
> Of any woman's bought complexion
> That the uprightest man—if such there be,
> That sin but seven times a day—broke custom
> And made up eight with looking after her.
> Oh she was able to ha' made a usurer's son
> Melt all his patrimony in a kiss. (1.1.20–27)

The skull's function as a sign for female sexuality becomes clear in act 3, when Vindice brings it on *"dressed up in tires"* (3.5.42.s.d.) to act as a decoy for the lustful Duke. Like Hippolito, we are unaware of Vindice's ruse, and in the theater we too should be shocked when, after some playful flirtation with the "lady," Vindice pulls back the veil. This literal exposure of the skull beneath the female skin is coupled with a metaphorical exposure as Vindice meditates on the ghastly object of his sexual passion:

> And now methinks I could e'en chide myself
> For doting on her beauty, though her death

Shall be revenged after no common action.
Does the silkworm expend her yellow labors
For thee? For thee does she undo herself?
Are lordships sold to maintain ladyships
For the poor benefit of a bewitching minute? (68–74)

Gloriana's skull becomes an image of Everywoman and a theme for one of
Vindice's great rhetorical set pieces.

   In spite of the wide difference in tone—from the Shakespearean sublime to
darkest Jacobean satire—Vindice's third-act address to the fleshless Gloriana
recalls Titus's third-act address to the tongueless Lavinia: Both emerge from
the same paradigm of gender and rebellion. The male appropriation of female
suffering, common to both, is neatly epitomized when the skull becomes a
weapon: Smeared with poison, it lures the Duke to his death. The device has
a witty propriety: The Duke's fatal kiss recapitulates the crime—his assault on
Gloriana—for which he is about to die. On another level, it exemplifies the
instrumental role of the martyred woman in the Lucrece myth.[5]

The republican tenor of the original Lucrece story posed an obvious problem
for its Jacobean adaptors. Heywood compensates for it by emphasizing
Tarquin's identity as usurper: His fall is occasioned as much by this original sin
as by his son's rape of Lucrece. In *The Revenger's Tragedy* the dramatist, unham-
pered by history, solves the problem in orthodox fashion: The revengers purge
the court of royal vice, but die for it (5.3.101–5). He nevertheless adheres to
the Lucrece pattern by having Lord Antonio—the Collatine figure—succeed to
the throne. As Vindice points out, the rape of Antonio's "good lady has been
'quited / With death on death" in the royal carnage (5.3.90–91), and Antonio
acknowledges heaven's justice (91) before sentencing the brothers to death.

   In *The Revenger's Tragedy* the good women are either dead, like Gloriana
and Lord Antonio's wife, or like Castiza, marginal to the main action. In either case they are remote, giving the male revengers a motive and a cue for
passion but taking no part in the larger drama. Symbolically, however, they
provide a foil for the lust-ridden world of the tyrant's court, and in this respect *The Revenger's Tragedy* foreshadows the later "Lucrece" variations.

Cymbeline

Shakespeare's *Cymbeline* was probably written soon after *The Rape of Lucrece*,
in 1608 or 1609,[6] and presents a sharp contrast to Heywood's play. It is an
experiment in form, a tragical-comical-pastoral-romance, written, like *The*

Revenger's Tragedy, for the King's Men and an audience more sophisticated than Heywood's. A "darkly comic study of imagined rape" (Stimpson "Shakespeare" 61), *Cymbeline* is erotic where *The Rape of Lucrece* is bawdy, and martial exploits play only a minor role. The heroine suffers neither death nor—permanent—dishonor, and the sexual threat she endures does not become a pretext for a rebellion. Nevertheless, Iachimo's specular violation of Imogen, which strongly recalls the rape in Shakespeare's *Lucrece,* points to larger affinities between *Cymbeline* and the other Lucrece/Virginia plays. Like her classical prototypes Imogen is persecuted by a lustful male (Cloten) who represents the corrupt government; her suffering proves redemptive for a diseased polity; and a final climactic battle marks the regeneration of the community. By connecting the wager plot derived from Boccaccio[7] with British history—that is, by making the wife of Boccaccio's tale the heiress presumptive to the British throne—Shakespeare invests the heroine's suffering with national, political significance.[8] Like Lavinia, Imogen embodies the state.

Imogen's marriage sets her apart from the other young heroines of Shakespeare's romances and connects her with Lucrece and Lavinia. As Granville-Barker remarks, "married chastity [. . .] is the chief theme of the play," and "Imogen is its exemplar" (2: 153). Her marriage is highly irregular, however—a secret contract quickly severed by paternal wrath—and her husband Posthumus implies that it may not have been consummated:

> Me of my lawful pleasure she restrain'd,
> And prayed me oft forbearance: did it with
> A pudency so rosy, the sweet view on't
> Might well have warm'd old Saturn; that I thought her
> As chaste as unsunn'd snow. (2.4.161–65)

The logic of the wager plot requires, of course, that Posthumus know Imogen's bedroom and the "cinque-spotted" mole. Like the time scheme of *Othello,* however, Imogen's sexual status is ambiguous.[9] The play directs us to view her as both wife and virgin, a duality reflected in the pictures of "Proud Cleopatra" and "Chaste Dian" which decorate her chamber (2.4.66–85). In her virginal aspect Imogen is—like Spenser's Amoret—an unravished bride, wedded but not bedded, and as such she is the focus of intense erotic excitement and anxiety.[10]

Four men, including her husband, father and stepbrother, persecute Imogen, and from the beginning the threat of sexual coercion looms. The

play's emphasis is thus far less on the opposition between good and evil political groups (tyrannical Tarquins/oppressed Romans), than on the extended
suffering of the heroine, upon whose integrity, literally and figuratively, the future of the commonwealth depends. The anonymous Second Lord draws our
attention both to Imogen's perilous position and her national significance:

> Alas poor princess,
> Thou divine Imogen, what thou endur'st,
> Betwixt a father by thy step-dame govern'd,
> A mother hourly coining plots, a wooer
> More hateful than the foul expulsion is
> Of thy dear husband, than that horrid act
> Of the divorce, he'ld make. The heavens hold firm
> The walls of thy dear honour, keep unshak'd
> That temple, thy fair mind, that thou mayst stand,
> T'enjoy thy banish'd lord and this great land! (2.1.55–64)

This summary also underlines the political parallel between Cymbeline's
court, Tarquin's and the Rome of the Andronici. In each case an ambitious
queen is the root of all evil; in each case that queen's son threatens the innocent heroine.[11]

The Second Lord's fervent prayer occurs ironically in the interval between Iachimo's failure to seduce Imogen (1.7) and his successful invasion of
her bedroom (2.2). The Lord's use of the traditional analogy between a
building and the female body ("the walls of thy dear honour") points to the
significance of the next scene. Immediately after his prayer we see Imogen in
her room, commending herself to the gods as she prepares to sleep
(2.2.8–10). Her prayer, unconsciously echoing the Second Lord's,[12] emphasizes her vulnerability. Iachimo's penetration of her room—figuratively, "the
walls" of her "dear honour"—follows directly.

Unlike his sources Shakespeare treats the bedroom scene erotically. Indeed he comes as close as he can to actually staging the rape he narrates in
Lucrece. It is, as many critics have noted, a theatrical tour-de-force. Unlike
the reader of Boccaccio or of *Frederyke of Jennen*, Shakespeare's audience is
ignorant of the villain's scheme. We are surprised by Iachimo's appearance
and probably apprehend a physical assault. Iachimo himself arouses expectations of a rape by comparing his movements to Tarquin's ("Our Tarquin
thus / Did softly press the rushes, ere he waken'd / The chastity he wounded"
[2.2.12–14]). Like Shakespeare's Tarquin, Iachimo hovers hungrily over the
sleeping woman, excited by her unconscious beauty.[13] The violation is more
than just specular: The text indicates a kiss ("That I might touch! / But kiss,
one kiss! Rubies unparagon'd, / How dearly they do't" [16–18]),[14] as well as

the symbolic theft of a bracelet. As with the parallel episode in *The Two Noble Ladies,* the audience's response to this scene will depend largely on the extent to which each member identifies with the passive woman and/or the active man. Shakespeare is again exploiting the peculiar excitement and/or horror that emerges from the encounter between male predator and unconscious female victim.

Imogen is not, of course, Lucrece; Iachimo does not "enjoy" her "dearest bodily part." Nevertheless, the difference between Tarquin's crime and Iachimo's is, in one sense, negligible: Iachimo has access to Imogen's body without her consent. Because he can prove this access publicly, he "steals" her honor, just as a rapist "steals" the chastity of the woman he rapes. In terms of the wager plot, the effect of an actual rape would be the same as that of the lesser, specular violation: Iachimo could prove his intimate knowledge of her body in the same way, and Imogen would be the same faithful wife and unconsenting victim. She would not, however, be physically *chaste* and, according to the conventions governing chastity in Jacobean drama, the consequences of this would be tragic: Imogen would have to die—either by her own hand or another's.[15] Because she remains technically pure, a comic ending is possible: With her "honor" restored to her, Imogen can resume her marriage with Posthumus. Her visual violation by Iachimo, like her subsequent mock death, is thus a tragicomic device: It is a rape that is not a rape. It allows Shakespeare to exploit theatrically the eroticism of *Lucrece* without sacrificing either his heroine or a happy ending.

The mock-rape also ensures, on one level, a blameless heroine. Since Imogen is asleep throughout the scene, she enjoys a manifest innocence that Lucrece has to prove by her suicide. Nevertheless this episode does affect our perception of her. As Neely observes, Iachimo's "salacious desires contaminate their object," while Imogen's "inert chastity, as he describes it, invites assault" (182). Imogen is degraded by Iachimo's objectifying gaze—in spite of the "reverence" of his courtly rhetoric—and, just as Shakespeare's verse shows us a Lucrece whose beauty makes her complicit in Tarquin's assault, so here Imogen's beauty is provocative. Two other small signs suggest that Imogen, on some level, shares responsibility for Iachimo's successful access to her body. In her bedtime prayer Imogen asks for protection from "fairies and tempters of the night" (9)—not from "all perils and dangers of this night," as *The Book of Common Prayer* puts it (64), but from *tempters.*[16] A tempter is "one who [. . .] tempts or entices to evil" (*OED* 1). This suggests that Iachimo comes to her in bed, as the devil came to the legendary Justina, as a figure of sexual temptation; that in fact he represents *her* desire. Secondly, Iachimo tells us that Imogen has been reading the "tale of Tereus" and that "the leaf's turn'd down / Where Philomel gave up" (45–46). Nosworthy glosses "gave up" as "yielded, succumbed" (51) and by this reading Iachimo implies Philomel's shared responsibility for her rape: She "gave up"—yielded or succumbed—as if to a pressing invitation to sexual pleasure. These two verbal signs—Imogen's fear of "tempters," and Iachimo's allusion to the yielding Philomel—frame the violation of Imogen and subtly suggest that on some level she shares Iachimo's illicit desires.[17]

Although Imogen's unconsciousness throughout the scene ensures her technical innocence, it also works against her. Unlike the legendary Justina, who wrestles with the devil, or even Shakespeare's Lucrece, who challenges Tarquin rhetorically, morally and emotionally, Imogen is entirely passive. She lacks the dignity conferred by action. She lacks even the dignity of consciousness. She is utterly unaware of what is happening to her, and this ignorance distances her from the audience. We know too much, she knows too little. Imogen becomes, for the length of the scene, merely a victim, and her victimization is our spectacle.

Iachimo's bedroom victory determines Imogen's fate. From this point she is subject to a progressive degradation that ends only with her rescue by Lucius (4.2). In his early morning visit—with its ironic aubade for her night with Iachimo—Cloten takes up the theme of assault in another key: Eroticism gives way to the crude bawdiness of his puns ("if you can penetrate her with your fingering, so: we'll try with tongue too" [2.3.13–14]). Iachimo's slanderous description of Imogen's "adultery" continues her degradation. Even though we know that Iachimo is lying about Imogen's surrender, his narrative powerfully creates—for us as well as for Posthumus—the thing it describes, her body:

> [. . .] under her breast
> (Worthy her pressing) lies a mole, right proud
> Of that most delicate lodging. By my life,
> I kiss'd it, and it gave me present hunger
> To feed again, though full. (2.4.134–38)

His speech constructs Imogen here, just as in her bedroom, as the object of his lust. Since in this second description he casts her as his willing partner, he also creates a false, promiscuous Imogen.[18] Posthumus, credulous and angry, creates in turn a grotesque caricature of this "Imogen," a pornographic cartoon in which she is reduced to an orifice for Iachimo's lust:

> O, all the devils!
> This yellow Iachimo, in an hour, was't not?
> Or less; at first? Perchance he spoke not, but
> Like a full-acorn'd boar, a German one,
> Cried "O!" and mounted [. . .]. (165–69)

Posthumus's initial cry of vengeance—"O, that I had her here, to tear her limb-meal! / I will go there and do't, i' th' court, before / Her father" (2.4.147–49)—foreshadows Cloten's plan to rape Imogen and then "knock her back" to her father's court (3.5.145). In both cases the imagined violence is conceived theatrically, as a public display of sexual dominance—in particular, as a demonstration of manhood to Cymbeline, who has been a father to both young men. In the event, Posthumus elects to murder Imogen by proxy. Clearly the death he orders is not simply the execution of an adulteress: It is a lust-murder, a form of sexual aggression.[19] As his surrogate, Cloten pursues Imogen dressed in Posthumus's clothes, acting out their shared desire for a violent sexual revenge.[20] Cloten also pays the penalty for Posthumus's signal lapse in faith. It is he, still dressed as Posthumus, who dies a violent death; and it is thus his body that Imogen mistakenly grieves over, clasping it in a bloody embrace (4.2.295–332).

This extraordinary misidentification has caused critics some difficulty. Granville-Barker calls it "dramatically inexcusable": "It is a fraud on Imogen; and we are accomplices in it. [. . .] Imogen herself is put, quite needlessly, quite heartlessly, on exhibition. How shall we sympathize with such futile suffering? And surely it is a faulty art that can so make sport of its creatures" (2: 178). More recently Michael Taylor, acknowledging our probable discomfort with this scene, has attempted to defend its dramatic propriety:

> however much [. . .] we would like to spare Imogen and ourselves her necrophilic embrace, we can hardly fail to notice that it seems in some respects no more than fitting that she should suffer such an indignity, the like of which would be unimaginable for Marina and Perdita [. . .]. Although we may flinch from its painful accumulation of detail, in a powerful way the indignity to Imogen satisfies expectations aroused in us during the course of Shakespeare's treatment of the wager story in *Cymbeline*, bringing to a suitably grotesque climax an element of punitive behaviour in relationships and towards the self [. . .]. (98)

Taylor is right to note the episode's propriety: It is, as he says, "a suitably grotesque climax" to Imogen's progressive humiliation. I believe he errs, however, in concluding that it is part of "a pattern of erotic punishment in which both lovers suffer for the naivety of their expectations" (105). This implies a symmetry to the lovers' experience. Their suffering, however, is not equivalent, nor equally deserved. Imogen suffers far more, and far more visibly, than Posthumus: Her pain is, as Granville-Barker puts it, "on exhibition." It is caused, moreover, not by her own "sexual frailty," as Taylor implies (105), but by her husband's. This attempt to rationalize Imogen's ordeal ignores the extent to which—in spite of the defiling eroticism of the "rape" scene, with its suggestion of her complicity—the play insists on Imogen's purity,[21] almost, indeed, her divinity. As Pisanio declares, "She's punish'd for her truth; and undergoes, / More goddess-like than wife-like, such assaults / As would take in some virtue" (3.2.7–9).

Like Lucrece and Virginia, Imogen is a holy victim, and like them bears a strong family resemblance to the virgin martyrs. In addition to being chaste, beautiful and wellborn, she is persecuted by her father for refusing a marriage, and punished for a sexual crime she did not commit.[22] Our sense of Imogen as beleaguered virgin is strengthened by a change Shakespeare makes to the traditional wager plot. While in the sources the villain does not even try to seduce the heroine, Iachimo does attempt a seduction—an attempt Imogen calls an "assault" (1.7.150). Like Virginia and Castiza, she repulses her assailant angrily, disdaining him "and the devil alike" (148).

Such language—the language of the morality plays—surrounds Imogen and her persecutors. To the Second Lord, the Queen is "a crafty devil" (2.1.51) and Imogen "divine" (2.1.56). For Iachimo in her chamber, Imogen's sanctity compels acknowledgment: "I lodge in fear," he cries, "Though this a heavenly angel, hell is here" (2.2.49–50). Indeed, Iachimo's emergence from the trunk at midnight strongly suggests a devil unloosed at the witching hour, as his fearful retreat on the stroke of three enacts the demon's return to the hell-mouth.

One of the most arresting religious images is Imogen's early description of Posthumus as her "supreme crown of grief" (1.7.4). As Michael Taylor points out (101), the lovers' situation scarcely explains this extravagant language. It is, however, prophetic of Imogen's approaching ordeal. It suggests the crown of thorns—a symbol of Christ's passion—as well as the crown he bestows on his brides, the Virgin Mary and the virgin martyrs. It is thus highly appropriate for Imogen, the virginal wife whose marriage becomes a species of martyrdom.

Imogen's spirited virtue prompts her vocal defiance of both her father (1.2.64–81) and Cloten (2.3.123–35), and her flight from the prison of Cymbeline's court. Once she learns that Posthumus believes her unfaithful, however, this spirit deserts her. She becomes a Patient Grissill, sanctified by her willingness to suffer. In contrast to the heroine of the source tales, who pleads with her husband's servant for her life, Imogen invites the death Posthumus has commanded. Indeed, she asks for execution repeatedly before Pisanio can unfold his plan to save her.[23] Religious imagery underlines her sacrificial role ("Prithee, dispatch," she urges, "The lamb entreats the butcher" [3.4.97–98]).[24] Though Pisanio prevents Imogen's physical death, the text makes it clear that she suffers a spiritual death here, symbolized by her subsequent burial. As Pisanio puts it, "What shall I need to draw my sword? the paper / Hath cut her throat already" (33–34).

At this point in the source stories the resourceful heroine, bent on survival, escapes in disguise, and by her efforts rises to a position of power and influence. Finally she organizes the triumphant denouement that restores her to her identity and honor. By contrast, the pathetic Imogen stumbles about the Welsh wilderness, at the mercy of the kindly inhabitants, the noble Romans and an apparently capricious Fortune that allows her first to be "buried" alive and then to mistake Cloten's headless body for Posthumus's. When at length the denouement does occur, it is, as the vision of Jupiter insists, in fulfilment of a divine, not human, design.[25]

In the microcosm of the royal family, Britain's decay is manifest in the persecution of Imogen, the heiress presumptive, and, on a national level, in the refusal of tribute to Rome. Symbolically Imogen's suffering and "death" atone for the sins of her family and country. There is also a direct practical connection between her ordeal and their rejuvenation. The first sign of that renewal is Guiderius's defeat of Cloten (4.2), and it is Imogen who unwittingly brings Cloten to Wales and his well-deserved death. Guiderius, equally unwitting, avenges the sexual threat to his sister and rids his country of a tyrant.[26] The second stage in Britain's recovery is the climactic battle that brings the princes out of hiding to save their father's life and kingdom.[27] Again, Imogen's martyrdom acts as a hidden cause of that victory, and thus of the final reunion: Posthumus, who helps the princes in their spectacular feat (5.3), is there to fight for Imogen and atone for her "murder":

> [. . .] 'tis enough
> That, Britain, I have kill'd thy mistress: peace,
> I'll give no wound to thee: therefore, good heavens,
> Hear patiently my purpose. I'll disrobe me
> Of these Italian weeds, and suit myself
> As does a Briton peasant: so I'll fight
> Against the part I come with: so I'll die
> For thee, O Imogen [. . .]. (5.1.19–26)

Posthumus's heroism is thus a form of religious service, an act of devotion to the wife he believes he killed. Finally, the recognition scene recapitulates Imogen's sacrificial role in the rejuvenation of her family/country: When she, as Fidele, attempts to comfort Posthumus, he strikes her angrily to the ground (5.5.229). Her fall prompts the chain of revelations that leads to the restoration of the princes and Britain's peace with Rome.[28]

If Heywood's *The Rape of Lucrece* and *Appius and Virginia* celebrate a political order in which women are finally irrelevant, *Cymbeline* offers a far more disturbing vision. Imogen's heroism is central to the play, but it is the heroism of the beaten wife. She is punished not—as Pisanio claims—for her truth, but for her *beauty,* her sexuality. Cruelly victimized, she "undergoes more goddess-like than wife-like" the assaults of her husband, who sends Iachimo to seduce her and Pisanio to kill her. The "divine Imogen" is the apotheosis of Patient Grissill.

The Second Maiden's Tragedy

The Second Maiden's Tragedy (1611), like *Cymbeline* and *The Revenger's Tragedy,* was written for the King's Men and has much in common with both.[29] In contrast to the complex intrigue of the earlier plays, however, *The Second Maiden's Tragedy* presents the Lucrece/Virginia theme with stark simplicity: There is only one tyrant, one victim, and one hero. (Their names— "the Tyrant," "the Lady," and "Govianus"—indicate their emblematic nature.) The action of the main plot is even simpler than that of Webster's play. Here the focus is primarily on the tyrant's extended sexual aggression, the victim's chaste response (suicide, posthumous supernatural soliciting) and the hero's revenge. The other nobles, who first support the Tyrant, then turn against him, play only a minor role.

As Lancashire has demonstrated, *The Second Maiden's Tragedy* is indebted to a variety of hagiographic and religious sources for its main plot; in particular the story of Sophronia—a Roman matron who committed suicide to escape the lust of the tyrant Maxentius—and also the Elizabethan versions of the Virginia story.[30] The Tyrant's necrophilia was certainly inspired by the legend of Herod and Mariamne, and possibly also by the apocryphal tale of Callimachus and Drusiana.[31] The main plot thus offers a celebration of chastity in conventional hagiographic terms. The carefully juxtaposed subplot reinforces the homiletic message: In a contrasting, domestic triangle, a woman succumbs to sexual temptation with fatal results.[32] As in *The Revenger's Tragedy,* the exceptional chastity of one woman proves the rule of female sexual weakness: Here the commonplace adultery of the Wife is a foil for the peerless chastity of the Lady.

Though much less complex than Imogen, the Lady shares her ambivalent sexual status. In acts 2 and 3 she is living with Govianus and described as his lady (2.1.s.d.), while Govianus is called the "lord unto so rare a wife" (4.4.24). As Lancashire points out (*Second Maiden's* 220), this could simply refer to a betrothal. The author leaves the exact nature of their bond vague, and thus—like Shakespeare—has it both ways: The Lady retains her virginity, with its powerful religious connotations, but also demonstrates her loving fidelity to a husband. Like Imogen, she is not simply chaste, she is a chaste *wife*.

Like Imogen, too, she demonstrates her worth by rejecting a marriage that is advantageous in worldly terms, and choosing, against her father's will, a morally superior, impoverished suitor. In *Cymbeline* this choice is made before the play begins; in *The Second Maiden's Tragedy* it is the pivotal event of the long first scene. The opening lines establish the respective positions of "*the new usurping* TYRANT" and "*the right heir* GOVIANUS *deposed*" (1.1.s.d.)— positions probably reinforced by the stage imagery: The Tyrant enthroned on high, surrounded by his supporters, and Govianus isolated below.[33] At line 10 the Tyrant introduces the theme of their rivalry for the absent Lady, at 26 she is sent for, and at 59 Govianus laments his inevitable loss of her ("O, she's a woman, and her eye will stand / Upon advancement, never weary yonder" [63–64]). Not until line 111, however, almost halfway through the scene, does the Lady enter. Then, "*clad in black*" (111.s.d.), she is the center of attention, watched by an audience onstage as well as off. Govianus's reaction signals her spectacular beauty: "Now I see my loss; / I never shall recover't. My mind's beggared" (111–12).[34] The Lady's black dress, ostentatiously modest, would present a vivid contrast to the brilliance of the Tyrant and his faction. She stands silently while the Tyrant protests her funereal attire (113–22), and then declares: "I am not to be altered" (123).[35]

These words, the first the Lady speaks, define her. She is essentially constant. In contrast to the dynamic power of Marina, her chastity is static. Immoveable in her virtue, the Lady is simply there, to be beheld. Her reply to the Tyrant—"I come not hither / To please the eye of glory, but of goodness" (127–28)—affirms the play's implicit definition of her as object of the male gaze. Though she eschews worldly splendor, her function is nevertheless specular. Moreover, whatever her intentions, she attracts the gaze of everyone, great *and* good. As the Tyrant complains, "where she goes / Her eye removes the court. What is he here/ Can spare a look?—they're all employed on her!" (147–49).

The Lady's primary act in this scene, her election of Govianus, is not fully enacted: It is not, that is, a deliberate choice that we watch. Like her black costume, it is a demonstration of a choice already made, a demonstration of her nature. Once the Lady has taken her place at Govianus's side—and she does this almost immediately—her role as an agent in the drama is all but over. As she puts it, "I have found my match, / And I will never loose him" (130–31): From this point on, she simply adheres to her spouse. She is, indeed, not even verbally active—speaking with wit, spirit and dignity, but little more than is necessary for the preservation of her chastity.[36] Thus in the second act, when approached by her father in his role as pander, the Lady—a model of filial decorum—firmly rejects his suit but allows Helvetius to speak, and at some length (2.1.20–41, 76–101). It is Govianus who responds actively, first threatening Helvetius with a pistol, then verbally assaulting him in an extended bout of moral invective (111–53). And it is thus Govianus, in spite of the Lady's saintly status, who works his conversion.[37]

Govianus remains the active partner—dispatching Sophonirus with grim humor (3.1.26–29), commanding his servants masterfully (3.1.6, 46, 53, 59–60)—until the Lady's imminent danger becomes clear. Then without faltering she takes the initiative, reproving Govianus for his hesitation and urging him to kill her:

> Come on, sir!
> Fall to your business; lay your hands about you.
> Do not think scorn to work. A resolute captain
> Will rather fling the treasure of his bark
> Into whales' throats than pirates should be gorged with't.
> Be not less man than he. Thou art master yet,
> And all's at thy disposing. Take thy time;
> Prevent mine enemy. Away with me;
> Let me no more be seen. (3.1.67–75)

Here and repeatedly in the lines that follow, the Lady presents her appeal for death in terms of gender and property: It is Govianus's right and duty as husband, lord and master to prevent her dishonor by killing her. "Sir, you do nothing," she cries, "there's no valour in you" (87); "I speak thy part, / Dull and forgetful man, and all to help thee!" (92–93). His reluctance not only heightens the pathos and suspense of this critical scene, it provides a foil for the Lady's heroic resolve. Like the virgin martyrs praised by the Church Fathers, like Lucrece, the Lady transcends the weakness of her sex and attains a "masculine" heroism.[38]

Significantly Govianus yields to the Lady's demands when she graphically depicts the alternative to her death: "Is it thy mind to have me seized upon / And borne with violence to the tyrant's bed, / There forced unto the lust of all his days?" (94–96). The thought of this violation, this sexual theft, is enough for Govianus: "O, no, thou liv'st no longer now I think on't. / I take thee at all hazard!" (97–98). His use of the verb "take" for "kill" underscores the sexual connotations of this killing. The Lady subsequently returns to the theme of sexual possession to strengthen his flagging resolution: "His lust may part me from thee, but death, never; / Thou canst not lose me there, for, dying thine, / Thou dost enjoy me still. Kings cannot rob thee" (144–46). Theologically, of course, the Lady has considerable authority for at least part of her statement: According to the Christian tradition enshrined in hagiographical legend, a victim of rape was in danger of eternal damnation, so the Tyrant's lust could indeed part the lovers. However, her assurance to Govianus that he will "enjoy" her eternally goes beyond this traditional fear of damnation by rape to suggest its inverse: the lovers' timeless erotic union achieved through the Lady's chaste death. In part this echoes the "bride of Christ" motif of hagiography, but its reference to an earthly spouse gives it a novel turn. In effect the mortal husband is elevated to the status of Christ and assured with religious zeal that his right of property in her will endure through eternity. When Govianus finally "*runs at her*" with his sword, the Lady hails him as her lover, "Never more dearly welcome" (149). However, his virility is temporarily eclipsed by a swoon, and the Lady cries in disappointment, "alas, sir! / My lord, my love!—O thou poor-spirited man!" (149–50). She is forced to appropriate his sword and kill herself, the "resolute lady" of a "fearful master" (158–59).

The playwright takes a risk in emasculating his hero at this critical moment, but gains a great deal by it. Primarily, of course, the Lady's suicide ensures her full heroic status as a martyr to chastity. Less obviously, but no less importantly, it masks the material significance of her death. In terms of this world it is Govianus who benefits—sexually, emotionally, politically—from the Lady's choice of death before dishonor: It is his, and ultimately the country's, interests that are served. However, the play disguises the disparity between the heroine's pain and the hero's profit by assigning full responsibility for the death to the Lady herself: It is her idea, her desire and, finally, her deed. Govianus has no part in it. In terms of the plot, the Lady's suicide relieves Govianus of any blood guilt: he can leap to his feet and, after his brief encomium to the Lady, resume his life, unshadowed by the remorse that haunts Webster's Virginius. His trick with the body of Sophonirus—propping it against the door so that the invading soldiers assume the blame (179–94)—resurrects Govianus in our eyes as the clever hero of the revenge tradition and signals his commitment to some form of struggle in this world.[39]

At this point in the classical Lucrece/Virginia paradigm, the hero uses the death of the heroine as a political rallying point. In dramatic terms her death would traditionally fall, as it does in the plays of Heywood and Webster, in the fourth act, and the community's regeneration in the fifth. *The Second Maiden's Tragedy* varies this pattern by presenting the Lady's death in act 3, then repeating—and magnifying—the Tyrant's sexual crime in act 4. Structurally the second crime is thus a surprise. The content of that crime— necrophilia—is even more novel. The conventional response of a tyrant thwarted in his lust is hatred and anger directed murderously toward the object of his former "love."[40] In his pursuit of the Lady beyond death this Tyrant is, to my knowledge, unique on the Jacobean stage.[41] His corpse infatuation, though drawn from legend, would show an audience a crime that was probably, for most of them, hitherto unimaginable.

When the Tyrant begins his preparations for the grave robbery (4.2.37), the audience, like the soldiers he commands, is unaware of his intentions. We watch him in the church (4.3), as we watch Iachimo in Imogen's bedroom, with horrified fascination. The two scenes are related through a partial inversion: Iachimo approaches a living woman and wishes her as senseless as a "monument / Thus in a chapel lying" (2.2.31–32); the Tyrant approaches a "monument" (4.3.9) in a chapel and wishes the dead woman inside it to live. In both cases the sexual predator pays homage to his victim in a perversion of religious worship. What is metaphorical in *Cymbeline,* however, becomes concrete here.

Repeated references to the silence of the place (1–4, 15, 24–25) would prompt an audience to hear the blows the Tyrant delivers to the tomb (37–49) as especially violent,[42] while the religious dread that seizes even the soldiers (27–56) heightens our sense of the Tyrant's sacrilege. If, as Lancashire suggests (*Second Maiden's* 208), the "monument" included a marble effigy of the Lady, then the Tyrant's assault on the tomb would provide an emblem of the sexual violation he intends. The scene culminates in his macabre courtship of the Lady's body, as he kisses and embraces it (84–122).

Like Cloten, Govianus, unaware of the violation his mistress has just suffered, arrives at dawn to pay an unintentionally ironic tribute; he too has an aubade sung to her.[43] Unlike Cloten, Govianus is rewarded almost immediately by his beloved's appearance: *"On a sudden, in a kind of noise like a wind, the doors*

clattering, the tombstone flies open, and a great light appears in the midst of the
tomb; his LADY, *as went out, standing before him all in white, stuck with jewels,*
and a great crucifix on her breast" (42.s.d.). The details of the Lady's spectacular apparition—the noise, the light, her costume—all suggest a powerful religious presence (Lancashire, *Second Maiden's* 222–23). However, the power
implicit in this glory is belied by the Lady's words as she reveals her plight and
lays upon Govianus the burden of restoring her body to the grave:

> I am now at court
> In his own private chamber. There he woos me
> And plies his suit to me with as serious pains
> As if the short flame of mortality
> Were lighted up again in my cold breast;
> Folds me within his arms and often sets
> A sinful kiss upon my senseless lip;
> Weeps when he see the paleness of my cheek,
> And will send privately for a hand of art
> That may dissemble life upon my face
> To please his lustful eye. [. . .]
> My rest is lost; thou must restore't again. (66–79)

Thus in spite of her transfiguration the Lady is more than ever confined to
the role of object, her spirit at the tomb pleasing Govianus's "eye of goodness," her body at the court pleasing the Tyrant's "eye of greatness."

The scene in the Tyrant's "private chamber," described by the Lady, depicts a rape that is not a rape, and here again the play parallels *Cymbeline.*
The Tyrant's "serious pains" suggest corporal penetration but do not specify
it: The precise extent of the Tyrant's crime remains vague. More importantly,
however, is that he is embracing a corpse, and a corpse may be violated but
not raped (a dead body cannot refuse consent). It is not really rape because
the body is not really the Lady. Like the unconscious Imogen, moreover, the
Lady cannot be blamed for the successful assault. Indeed, since she is *dead*
her spirit can report the violation of her body without having experiencing
it. Her lip is "senseless," her breast "cold:" She feels nothing and is manifestly
innocent of the carnal pleasure that raped women are supposed to enjoy. In
a sense this necrophilia is only the logical outcome of the drama's dialogue
with hagiography: If a playwright wanted to have an *unequivocally* chaste
heroine actually suffer violation, he would have to arrive at the solution of
The Second Maiden's Tragedy—first death, then dishonor.

●●●

In act 5 the Lady again functions prominently as spectacle. Her body is
brought in a chair, *"dressed up in black velvet which sets out the paleness of the*

hands and face, and a fair chain of pearl 'cross her breast, and the crucifix above it" (5.2.13.s.d.). The stage directions insist on the aesthetic impact of the Lady's costume: The black heightens the pallor of her skin, and makes her perhaps more beautiful to the Jacobean eye, certainly more corpse-like.[44] Like Gloriana's skull dressed up by Vindice in fashionable clothes, the Lady's body would present an image of death-in-life; and just as in *The Revenger's Tragedy* the skull becomes the occasion for a meditation on sensual folly, so here the Lady's body, reverenced by the Tyrant, becomes a moral emblem. The song that accompanies his devotion provides, like Vindice's tirade, a moral comment. Again, there is a misogynistic subtext to this morality:

> O, what is beauty, that's so much adored?
> A flatt'ring glass that cozens her beholders.
> One night of death makes it look pale and horrid;
> The dainty preserved flesh, how soon it moulders.
> To love it living it bewitcheth many,
> But after life is seldom heard of any. (14–19)

As in *The Revenger's Tragedy,* so here the hero kills the tyrant by poisoning the lips of the dead heroine, and here too the device has a moral propriety, as the tyrant is poisoned through his own lust. Again, however, the trick epitomizes the hero's manipulation of a woman's death for a political end. In *The Revenger's Tragedy* that manipulation is blatant, underlined by Vindice's delight in his own wit. Here it is disguised by the religious burden the Lady imposes upon Govianus: As Lancashire notes (*Second Maiden's* 225), it is the burden not of vengeance, but of restoring the Lady's body to its rest (4.4.79). Instructed by her (5.1.200), Govianus poisons the Tyrant, then declares his readiness to die for the deed (5.2.141–51). His acclamation as king by the rest of the nobles takes him by surprise (179–80). Innocent of any overt political designs, he is nevertheless rewarded politically. At the play's close, Govianus regains his kingdom in this world, and the Lady, the "queen of silence" (205), resumes her reign in the next.

The Second Maiden's Tragedy represents the acme of the religious mystification of chastity on the Jacobean stage. The Lady combines the dignity of Lucrece and the purity of Virginia with the sanctity of the legendary Sophronia. Endowed with more apparent power than any of the other heroines treated in this chapter, the Lady devotes it wholly to maintaining her chaste bond with Govianus. It is she, not he, who insists on death before dishonor, and it is she who prompts the killing of the Tyrant. Symbolically the Lady's sacrificial death purges the corrupt court; practically her assault by the Tyrant—both before and after death—moves the nobles to rebellion (4.2.61–71). Her spirit prompts the Tyrant's death, and she is responsible for the nearly bloodless revolution that restores Govianus to the throne. No wonder then that the dramatist offers her as an exemplum of her sex. The final lines of the play, spoken by Govianus, are directed to the women in the audience: "I would those ladies that fill honor's rooms / Might all be borne so honest to their tombs" (5.2.211–12).

Valentinian

In *Valentinian* (c. 1612)[45] Fletcher uses the tyrant's pursuit of the heroine conventionally, for suspense and melodrama, and her voluntary death for pathos. However, he also reworks the classical paradigm in ways that expose the tensions inherent in it: In marked contrast to *The Second Maiden's Tragedy, Valentinian* reveals not only the husband's material interest in the victim's death, but the pressure on the assaulted woman to oblige him by dying. Even more prominent, however, is the disruption of the husband's homosocial community by the wife's rape and death. (As Machiavelli observes, "it appears that women have caused much destruction [. . .].") Like *Appius and Virginia, Valentinian* glorifies male bonds, and here too the heroine's suffering is eclipsed by that of a noble soldier. Fletcher goes further than Webster, however: His soldier is literally martyred, and it is his death, not the heroine's, that provokes the rebellion.

Like *The Second Maiden's Tragedy, Valentinian* draws upon one of the more sensational episodes in the history of the Late Roman Empire. According to Procopius, the depraved Valentinian—a fifth-century emperor—raped the wife of one of his nobles, Petronius Maximus. In revenge Maximus murdered the emperor, after arranging the death of a general whose opposition he feared. Proclaimed emperor, Maximus compelled the widowed empress to marry him, and she, to avenge her husband's death, summoned the Vandals to overthrow him.[46] Fletcher wisely omits the Vandals—his empress simply poisons Maximus at his coronation feast (5.8.52–64)—but adds a major complication: an intimate friendship between Maximus and the general Aecius, an affective bond far more intense than that between Maximus and his wife.

During the first two acts the friends—both soldiers—diagnose Rome's ills and debate the limits of obedience owed to the tyrant Valentinian. In counterpoint Fletcher provides an illustration of Rome's decadence. We see the lustful emperor—aided by a large support staff—pursuing the heroine. Like the Lady in *The Second Maiden's Tragedy* this paragon is praised for some time before she enters: beautiful, chaste and not yet 18 (1.2.13–20), she has, like the Lady, rejected the temptations of wealth, power and pleasure (32–67). Her name, Lucina, suggests a combination of "Lucrece" and "Marina," and her behavior confirms it. Fletcher indeed doubles the number of Marina's tormentors: four panders and two bawds beset Lucina to no avail (2.1–2). Like Marina she challenges the bawds' identity as women (1.2.27), calling them "Devills" (41) and attempting to exorcize them (43–47). Here, as in *The Revenger's Tragedy* and *The Second Maiden's Tragedy*, the exceptional woman proves the rule of female corruption. Nevertheless, by the end of act 2 Valentinian has trapped Lucina in his palace and led her off to her violation.

Fletcher's treatment of the rape itself is thus highly conventional. In his handling of its consequences, however, he departs from convention. He is the first playwright to dramatize a confrontation between rapist and victim after the crime,[47] and he plays on the contrast between Lucina's passion and the emperor's indifference. Initially this rape victim seems strong enough to threaten her assailant. Reproaching Valentinian with the ferocity of Philomela,[48] she promises to expose him ("As long as there is motion in my body, / And life to give me words, Ile cry for justice" [3.1.32–33]), and prophesies his ruin in apocalyptic terms ("the Gods will find thee, / [. . .] for they are righteous, / Vengeance and horror circle thee" [41–43]). However—again, like Philomela—Lucina invites her ravisher to kill her, and laments the stain on her soul:

> [. . .] flye from me,
> Or for thy safety sake and wisdome kill me,
> For I am worse then thou art; thou mayst pray,
> And so recover grace; I am lost for ever,
> And if thou lets't me live, th'art lost thy selfe too. (64–68)

More secure than Tereus, Valentinian only mocks her threats (34, 69, 118–31). Lucina, however, is no Philomel. Like Heywood's and Shakespeare's Lucrece she has internalized the ideology of chastity too wholly to survive her rape. Like them she feels herself monstrous, an outcast, deprived of her identity:

> I am now no wife for *Maximus,*
> No company for women that are vertuous,
> No familie I now can claime, nor Country,
> Nor name, but *Cesars* Whore. (74–77)

This sense of shame and guilt prevents Lucina from seeking refuge or—in spite of her threats—justice at home. She simply collapses (148), and when

Maximus and Aecius discover her (150) she does not demand revenge, or even tell them her story.

In a sense Lucina does not need to. Maximus surmises the truth almost instantly, and his response is fascinating: Dismissing his wife ("Rise, and goe home"), he turns at once to Aecius, "I have my feares *Aecius:* / Oh my best friend, I am ruind" (155–56). Maximus thus understands his wife's rape as an attack on *himself,* as *his* "ruin," and of course, he is in a sense correct. The emperor's theft of his sexual property has placed Maximus in an intolerable position: He either has to seek revenge or accept his position as cuckold to Valentinian. His first anxiety, however, is to ensure Lucina's suicide: "[G]o silver Swan,/ And sing thine owne sad requiem:[49]/ goe *Lucina,* And if thou dar'st, out live this wrong" (155–61). Lucina, of course, replies, "I dare not" (161). According to Livy (and Shakespeare, and Heywood), when Lucrece tells her story, the men declare her innocent and try to comfort her: Her suicide is a shock to them (bk. 1, ch. 59). Here the victim's husband does not even make this gesture: Maximus simply dispatches her to take her own life, according to a code that she accepts without question. She just says pathetically, "Long farewell Sir. / And as I have been loyall, Gods think on me" (186–87).

Fletcher draws out the farewell for some hundred and twenty lines, however, partly for pathos, partly for suspense. Aecius asks Lucina to live "a little longer," "a short yeare" (207–8) to bring the emperor to repentance: "For who knowes but the sight of you, presenting/ His swolne sins at the full, and your faire vertues, / May like a fearefull vision fright his follies" (212–15). He seems to imply that she should live to bear Valentinian's child, in a second, extended martyrdom. (An argument, one would think, unlikely to succeed.) He also urges her innocence, indeed, her sanctification through the violence she has suffered (220–31). Maximus argues against him, for the sake of his progeny:

> [. . .] could the wrong be hers alone, or mine,
> Or both our wrongs, not tide to after issues,
> Not borne a new in all our names and kindreds,
> I would desire her live, nay more, compell her:
> But since [. . .] our names must find it,
> Even those to come; and when they read, she livd,
> Must they not aske how often she was ravished,
> And make a doubt she lov'd that more then Wedlock?
> Therefore she must not live. (235–45)

Lucina settles the dispute by insisting on death (247–62), and both men pay a sentimental tribute to her chastity as she leaves (263–80).

The scene is striking because the male anxieties about rape, voiced only by the *heroine* in the Lucrece story, are here acknowledged by the husband. His interest in Lucina's death is explicit and inexorable. It is, moreover, grotesquely juxtaposed with the mystifying rhetoric of his eulogies ("Thou starry vertue, farethee-well; seeke heaven, / And there by *Cassiopeia* shine in glory, / We are too base and dirty to preserve thee" [190–92]). One might argue that Fletcher exposes this interest to challenge it, that we are prompted to condemn Maximus for his selfishness. This seems improbable, however: Dramatically, Fletcher himself has too much at stake in the ideology of chastity to challenge its rationale.

In Procopius the (nameless) wife goes home, curses Maximus for allowing the emperor access to her, and dies sometime later of unexplained causes (Loeb edition 2: 41, 45). In d'Urfé's *Astrée*—Fletcher's second source—the heroine, like Lucrece, exacts a promise of revenge from her husband, who dissuades her from suicide (2: 521–22); she lives to nag him on to his revenge (532), finally expiring with joy after she washes her hands in Valentinian's blood.[50] More fragile and passive than either of these source figures—and perhaps, therefore, more attractive to a Jacobean audience—Lucina simply dies of shame on the threshold of her house ("Dare I, said she, defile this house with whore, / In which his noble family has flourish'd? / At which she fel, and stird no more" [3.1.366–68]). Fletcher thus rewrites Lucrece's death, purging it of "masculine" valour. Lucina may be pitied and admired by the men, but not emulated. Her heroism poses no moral challenge.

With Lucina out of the way in 3.1, Fletcher can develop his major interest: The conflict between Maximus's desire for revenge and his love for Aecius, who opposes that revenge. Initially, Maximus expresses his dilemma as a choice between his wife and his best friend, and decides in favor of the latter: "Is he not more to me, then wife? then Cesar, / [. . .] Is he not more then honor, and his freindship / Sweeter than the love of women?" (3.3.22–25). He determines to forget Lucina: "let her perish. / A freind is more then all the world" (34–35). However, he changes his mind when he reflects that *other men* may say he was dissuaded from revenge by "fondnes of a freindship" (41–46). Maximus's need to be perceived as virile by men in general finally outweighs his devotion to one particular man.

Again, the scene is interesting because it makes explicit what is only implicit in other plays. The crucial relationships here, as in Heywood and Webster, are the male ones. As Maximus's deliberations make clear, Lucina counts for little in herself, but because he was her husband his masculinity is at stake in avenging her ("Was she not yours? Did she not dye to tell ye / She was a ravish'd woman?" [38–39]). Maximus's role as injured husband overrides the claims of friendship and country, and he ignobly betrays both.[51]

In Heywood's *The Rape of Lucrece* and *Appius and Virginia,* the death of the heroine strengthens the male bonds among her kin and their allies: Although Webster introduces tension between Virginius and Icilius, in both plays the (male) community unites to support the injured (male) party, and punish the (male) predator. In *Valentinian,* however, the woman's death disrupts the male bonds: The community—represented by Aecius—refuses to support the injured man. It is thus Maximus's "anxious masculinity," his investment in his wife's chastity, that alienates him from Aecius—"that pretious life I love most" (3.3.8)—and his community. (Significantly Aecius, the epitome of masculine virtue, has no wife or mistress; "I am too course for Ladies," he declares [1.3.115]). The ideal implicit in Fletcher's treatment of gender seems to be the exclusively male homosocial world of the army (the context of the friendship between Aecius and Maximus), and this is congruent with his portrait of Rome's decadence under Valentinian. Historically, as Fletcher would have known from Procopius, it was a period of intense military activity.[52] He represents it, however, as a time of slothful peace imposed upon the army by the sensual emperor. "Baudes, and singing Girles" steer the empire while the "glory of a Souldier" decays (1.3.7–9). The soldiers chafe at their confinement in Italy, where their "weapons / And bodyes that were made for shining brasse, / Are both unedg'd and old with ease, and women" (183–85). They long for battle and "tell their wounds, / Even weeping ripe they were no more nor deeper, / And glory in those scarrs that make 'em lovely" (192–94).

The emotional climax of the play is Aecius's spectacular death in 4.4—a process Fletcher extends for a grueling 270 lines—and it is a scene of intense male bonding.[53] Falsely accused by Maximus, Aecius has been sentenced to death by Valentinian. Though he is urged by his loyal servants to escape and raise a rebellion, Aecius scornfully refuses (4–24). Three panders sent to execute him cower in terror, though he offers no resistance (his martial demeanor alone is enough to frighten them) (93–118). Finally, Pontius—a noble officer whom Aecius had earlier discharged on suspicion of treason (2.3)—arrives and prepares to carry out the sentence. Instead of killing Aecius, however, he kills himself in a theatrical demonstration of loyalty to the general and to Rome. Fletcher gives him 33 lines of impassioned rhetoric (179–212) before Aecius breaks in to express his admiration:

> I want a name to give thy vertue Souldier,
> For only *good* is farre below thee *Pontius*

> [. . .] thou hast fashiond death,
> In such an excellent, and beauteous manner,
> I wonder men can live. (213–17)

With his last breath (almost) Pontius begs for the general's hand and his forgiveness, then dies regretting that he cannot inflict more wounds on himself
(226). Aecius is enraptured:

> Is there an houre of goodnesse beyond this?
> Or any man would out-live such a dying?
> Would *Cesar* double all my honors on me,
> And stick me ore with favours, like a Mistris;
> Yet would I grow to this man: I have loved,
> But never doated on a face till now:
> O death thou art more than beautie, and thy pleasure
> Beyond posterity: Come friends and kill me; [. . .]
> Come, and Ile kisse your weapons. (227–37)

Aecius, however, can find no partners for his orgy. He has to kill himself, in
an autoerotic frenzy: "hold my good sword, / Thou hast been kept from
bloud too long: Ile kisse thee, / For thou art more then friend now [. . .]"
(250–53).[54]

Though neither Procopius nor d'Urfé record an armed rebellion, Fletcher
adheres to the Lucrece paradigm. It is, however, *Aecius*'s death, not Lucina's,
that sparks the insurrection. Two of the general's servants, prompted by
Maximus, poison the emperor (5.1) and the army rises in revolt "for great
Aecius" (5.2.154). (Lucina does play a token role, however: We hear that
the "women of the Towne have murderd / [. . .] Caesars she-Bawdes"
[157–58], presumably in reparation for her death.) Maximus rides to power
on the "sluce of blood" he has opened (5.3.1) and immediately meets his
well-deserved end. The concluding lines of the play express a pious wish for
Rome's purgation: "Take up the body, nobly to his urne, / And may our
sinnes, and his together burne" (5.8.120–21).[55]

In Fletcher's revision of the classical pattern, Aecius, not Lucina, embodies redemptive virtue: In his fiercely active suicide he combines the martyrdom of Lucrece with the heroism of Brutus. Fletcher thus goes further than
Heywood and Webster in celebrating a world in which women are marginalized: In *Valentinian* the wounded male body is the site of virtue and the
focus of erotic pleasure.

All's Lost By Lust

At the end of *Valentinian,* as at the conclusion of all the earlier "Lucrece" plays, we have a sense of purgation, and the possibility of a new beginning. However, if Fletcher had followed his sources, instead of adhering to the Lucrece paradigm, the play would have ended with the Vandals plundering Rome. Rowley's *All's Lost By Lust* (1619–20)[56] dramatizes just such an ironic Lucrece analogue. The heroine survives her rape and provokes her father to a rebellion that enslaves rather than liberates her country. She is a failed Lucrece, he a failed Virginius. Though Jacinta is wholly innocent of any complicity in her assault, her failure to commit suicide after the event, her virulent anger and her demand for revenge against a legitimate king, place her beyond the pale of Jacobean sympathies. Like Ovid's Medusa, Jacinta becomes demonic through sexual violation.[57]

The play draws powerfully on two legends surrounding the Islamic conquest of Spain in the eighth century. According to one story, each Spanish king in succession had placed a padlock upon the door of an enchanted temple (the *casa de Hércules*). At his accession Roderick, the last Christian king, refused to follow custom and impiously broke into the building, where he found pictures of the Arabs and a letter saying, "When this door shall be opened, these people will invade this country" (Abd al-Hakam 20).[58] According to a second, more popular, tradition Roderick precipitated the invasion when he raped the daughter of a powerful noble, Julian, who, to avenge his honor, betrayed Spain to the Moors.[59] The French historian Lévi-Provencal observes that

> the unfortunate girl had to bear responsibility for all the evils that descended upon Spain from the day the country fell to the Moors. A whole literature would be inspired by the daughter of Count Julian: numerous ballads, of a much later date, and poems of the Romancero tell how she was seen by Roderick while bathing in the Tajo, at Toledo. They styled her by the name of Florinda and the infamous nick-name of "Caba" or "Cava" (from the Arab word meaning "whore").[60]

The girl is thus traditionally blamed both for her own rape and the subsequent Moorish "rape" of Spain—a blame epitomized in the scornful epithet, "la Cava."[61] The two legends of the enchanted tower and Julian's daughter are parallel ways of accounting for the same event, but Rowley united the stories, and in his well-crafted play Roderick's violation of the enchanted tower (called a castle) follows and recapitulates his violation of the girl (called Jacinta).

Although Jacinta is never called "la Cava" in the play, Rowley seems to have known the Arabic name: He joins the two legends through a play on the Spanish *cava*. Related to the verb *cavar*, to dig (from the Latin *cavare*, to hollow out), in Spanish *cava* refers specifically to the vault under a royal palace used for storing water and wine (*Diccionario de la Lengua Española*, *cava* 2.1). This sense of an underground vault was certainly in Rowley's mind: His Rodericke[62] speaks as if the center of the enchanted castle—"the fatall chamber" (5.2.1)—is in fact subterranean (1.1.48–50, 66–67). Thus the king's rape of the girl (la Cava) is reenacted in his forcible entry of the vault (la cava).[63]

The first scene swiftly juxtaposes the king's private resolution to force Jacinta (14–18), the news of the Moors' impending invasion of Spain (19–21) and Rodericke's public resolution to break into the forbidden castle (48–55): Virgin, castle and country are thus linked by an implicit analogy. By implication too the lustful king is associated with the infidel invaders. "Sooty as the inhabitants of hell" (33), the Moors on one level represent the demonic forces within the king and his corrupt court—forces that are already preying upon the country.

In contrast to the legendary depiction of Julian's daughter as "la Cava," Rowley's treatment of her conforms, initially, to the hagiographic conventions of the Jacobean stage. Like Castiza, the Lady, Virginia and Lucina, Jacinta is a virgin martyr pursued by a lustful tyrant, and like them she is impervious to the solicitations of the tyrant's pander. Thus Lothario reports to his master:

> She leanes another way, and talkes all of the spirit.
> I frighted her with spirits too, but all
> Would not doe: she drew her knife, pointed it to
> Her breast, swore she would doe something [. . .]. (1.1.9–12)[64]

By act 2 Jacinta—like Lucina—is trapped in the tyrant's palace, besieged by an evil bawd. (The bawd has instructions to "blow up" "the maidenhead [. . .] / In this enchanted castle" [2.1.29–30]). Like Marina, Jacinta engages in virginal fencing (2.1.38–69) as well as righteous indignation ("Out, shame of women! thou the falsest art. [. . .]" [77–85]). Jacinta, however, fails to save her chastity: Like Tarquin and Valentinian, Rodericke is inexorable.

Everything about the prelude to this rape is conventional—it strongly recalls Shakespeare's *Lucrece*—until Jacinta goes beyond rhetoric, tears and passive prayers to an active curse: "See, heaven, a wicked king, lust staynes his crowne; / Or strike me dead, or throw a vengeance downe" (2.1.147–48). Rodericke scoffs, "Tush, heaven is deafe, and hell laughs at thy crye," but Jacinta repeats her malediction ("Be cursed in the act, and cursed dye" [150]), before the king drags her off (151.s.d.). For a contemporary audience Jacinta's formal imprecation, probably delivered on her knees, would be an impressive verbal strike against the king, one rendered all the more powerful by Rodericke's guilt. Keith Thomas observes that "although post-Reformation Protestants usually denied both the propriety and efficacy of ritual cursing, they frequently believed that, if the injury which provoked the curse were heinous enough, the Almighty would lend it his endorsement" (*Religion* 605). Paradoxically, however, "[i]n the writings of the demonologists, as in the prosecutions before the courts, successful cursing and banning was treated as a strong presumption of witchcraft" (*Religion* 611). The curse of the injured woman was thus an ambivalent weapon: Even though she invoked God's help, the harm that followed was generally ascribed to the devil (*Religion* 611). By calling down divine vengeance on Rodericke, Jacinta places herself in the company of witches, beyond the pale of the early modern community.

None of the earlier heroines curses the tyrant before the crime.[65] Jacinta's indictment of Rodericke foreshadows not only his doom, but also her unconventional response to the rape. Before the event Rowley prompts us to expect Jacinta's suicide. We have Jacinta's own Lucrece-like gesture, reported by Lothario and quoted above ("She drew her knife [. . .]"), and we have her father's explicit challenge to her as he leaves for war in act 1. When Jacinta tells him of Rodericke's pursuit, he dismisses her fears angrily as treason (1.3.85–96); she cries, "There has bin ravishers, remember Tarquin," and he responds, "There has bin chast ladies, remember Lucres" (97–98). However, in spite of this strong and authoritative patriarchal admonition to suicide, Jacinta departs radically from the approved model after her rape. Unlike Lucrece (both Shakespeare's and Heywood's) and Lucina, she is not in the least self-destructive: she is just *angry*.[66] While for modern readers Jacinta's unalloyed anger is no doubt a welcome contrast to the suicidal grief of Lucrece and Lucina, for Jacobean audiences it would probably have compromised her status as a sympathetic victim.[67] As Lamb observes, the "striking absence" of anger in most Renaissance representations of Philomela "suggests the anxiety elicited by women's anger within Renaissance culture" (218).[68] The threatening aspect of the earlier raped heroines—Lavinia, Lucrece and Lucina—is in each case far outweighed by pathos. The first has no tongue or hands to threaten anyone, while the violence the others direct against themselves neutralizes the potential danger of their anger. Jacinta's failure to take the guilt of the rape upon herself makes her much less pathetic and far more threatening than her predecessors.[69]

Jacinta's transformation from saintly victim to avenging fury, signaled by her curse upon the king in act 2, continues immediately after the rape when she launches a verbal assault on her jailer, Lothario.[70] Threatening to "waken heaven and earth" and "astonish hell for feare" (3.1.5–6), she reviles him at length and with remarkable fluency:

> O that I could spit out the spiders bladder,
> Or the toads intrals into thee, to take part
> And mixe with the diseases that thou bear'st,
> And altogether choke thee! or that my tongue
> Were pointed with a fiery pyramis
> To strike thee through, thou bundle of diseases! (19–24)

Jacinta even spits at him (17). Her conjuring of "spiders bladder" and "toads entrals" here echoes the language of Jacobean stage witchcraft.[71] Given the widespread belief in the performative power of language in this period, Jacinta's verbal malevolence—like her earlier cursing of Rodericke—characterizes her as dangerously witch-like. According to Dolan, "[i]f anger was the source of witches' power, and sometimes feared as a kind of agency in itself, speech was the primary means of expressing that anger, of provoking the devil, and of enacting ill will. For witches, speech was effective, that is, destructive, action" (198). Jacinta's curses also suggests an internal corruption, as if her physical violation has infected her morally: She perceives herself, as well as Lothario, to be a source of disease. While before the rape Jacinta called only on heaven for help, after the rape she resolves to tell her "sorrowes unto heaven" and her "curse to hell" (3.1.61–62)—a division that suggests her newly ambivalent nature: part martyr, part fury.[72]

Jacinta's anger makes her dangerous not only to Rodericke but to her community. Indeed, she identifies king with country ("Spaine has dishonour'd and imprisoned me" [52–54]). Unlike Tarquin and Govianus, Rodericke is not a usurper. He is Spain's rightful ruler, and Jacinta's pursuit of vengeance has tragic consequences for the whole country. Ironically Jacinta foreshadows the self-destructive outcome of her quest when she cries to Lothario:

> O that thine eyes were worth the plucking out!
> Or thy base heart the labor I should take
> In rending up thy bosome! I should but ope
> A vault to poyson me, detested wretch [. . .]. (3.1.41–44)

In insisting on revenge Jacinta does open "a vault" that poisons her, just as Rodericke in raping her—as in violating "the fatall chamber" under the castle—opens a vault that poisons him.[73]

❋ ❋ ❋

Like the "fowle-faire" Jacinta (4.1.95), the enchanted castle—her objective correlative—has an ambiguous status, powerful for good or ill. In the first act Julianus warns the king that diviners

> [. . .] with fearefull prophesies predict
> Fatall events to Spaine, when that shall be
> Broke up by violence, till fate hath runne
> Her owne wasting period; which out staide
> Auspitiously, they promise, that wreathes are kept
> In the fore-dooming Court of destiny,
> To binde us ever in a happy conquest. (1.1.58–64)

If the castle's sanctity is respected it is beneficent, if violated, it will bring disaster. This ambiguity makes it an appropriate emblem for Jacinta's virgin sexuality: The sacred rites of marriage would ensure its "happy conquest";
"broke up by violence," it brings Rodericke's destruction.[74] Significantly it is
Julianus, Jacinta's father, who urges Rodericke not to violate the temple.
With Macbeth-like bravado, however, Rodericke defies Julian's warning:
"Tut, feare frights us not, nor shall hope foole us. / If neede provoke, wee'le
dig supply through hell / And her enchantments" (65–67).

Jacinta's aphorism on lust ("Can lust be cal'd love, then let men seeke hell,
/ For there that fiery diety doth dwell" [2.1.118–119]) foreshadows the apocalyptic scene in the last act when Rodericke, metaphorically reenacting the
rape, seeks a literal hell. By this time he has begun to experience the moral and
political consequences of his crime and Jacinta's curse: "[T]here's within my
bosome / An army of furies mustered, worse than those / Which follow
Julianus: conscience beats / The drum of horror up" (5.1.10–13). Nevertheless, Rodericke resolves to violate the castle: "Our selfe will search / And breake
those dangerous doores which have so long / Kept Spaine in childish ignorance" (18–20). His determination recalls the original sin of Adam and Eve—
the rebellious seizure of knowledge—as well as his own particular sin, the rape
of Jacinta.[75] Amid *"Thunder and lightening"* (5.2.s.d.), he enters the "fatall
chamber" (1) and sees there a show of "Devils" (4) who impersonate his submission to the invaders, including "that she-curse," Jacinta (18). What Rodericke confronts are thus, literally, the consequences of the rape. The enchanted
castle has become a hell-castle, and an image of his fury-ridden "bosome."[76]

Like Brutus and Virginius, Julianus appeals to his countrymen to join him
in revenge, and they agree unanimously (172–73). He goes beyond the classical precedent, however, when he frees his Moorish captives and invites
them to march with him (174–78). Although none of the Spanish nobles
protest this invitation, its disastrous outcome is foreshadowed by Prince Mulymumen's immediate petition to marry Julianus's "ravisht daughter" (180).
Jacinta's horrified response—"O my second hell, / A Christians arms embrace an infidell" (182–83)—suggests that like her, Spain is threatened with

violation by a second and worse tyrant, whose racial difference is figured by sexual aggression and monstrous cruelty.[77] In the final scene the victorious Mulymumen orders Julianus's eyes and Jacinta's "exclaiming tongue" plucked out (39). The eloquent Jacinta is thus punished, like Philomela, for her complaints, and her father, like Shakespeare's Gloucester, suffers for his moral blindness. When they reappear, mutilated—the speechless daughter leading the blind father—Julianus acknowledges the justice of their punishment: "This worke / Is none of thine," he tells Mulymumen,

> tis heavens mercifull justice;
> For thou art but the executioner,
> The master hangman, and those ministers,
> That did these bloody ravishments upon's,
> Thy second slaves. And yet I more deserve,
> I was a traytor to my lawfull king;
> And tho my wrongs encited on my rage,
> I had no warrant signde for my revenge.
> Tis the peoples sinnes that makes tyrants kings,
> And such was mine for thee. (134–44)

In a final punitive gesture, Mulymumen tricks Julianus into slaying his daughter (151–80). Jacinta's death is thus a grotesque parody of Virginia's and Lavinia's. While Virginia's is chosen by herself, and Lavinia's deliberately wrought by a loving father, Jacinta's is just a cruel joke.

❋ ❋ ❋

In contrast to all the earlier "Lucrece" plays, there is no sense of renewal at the close of *All's Lost By Lust*. Rowley represents Mulymumen as an archvillain, an anti-Christ, and his triumphal exclamation—"Let chroniclers write, here we begin our raigne, / The first of Moores that ere was king of Spaine" (203–4)—ends the play with a somber vista of Christian servitude to the infidel.[78] In spite of her initial transformation from "la Cava" of legend to stage virgin martyr, Jacinta shares the guilt for the fall of Spain. Her angry eloquence, so attractive to the modern reader, is demonized by Rowley: Jacinta becomes a "furie" in a "fowle-faire shape" (4.1.95) whose "exclaiming tongue" precipitates a disastrous rebellion. Like Medusa—like Lavinia, Lucrece and Lucina—Rowley's Jacinta is thus made monstrous through her rape.

Hengist, King of Kent

Middleton's *Hengist, King of Kent* (1619–20)[79] stands apart from all the other "Lucrece" plays. Here the classical paradigm is subject to a final ironic transformation: The figures of the lustful tyrant and the husband are united, the heroine's suffering is mocked, and the male heroism of the last act is purely cursory. If Middleton did indeed write *The Revenger's Tragedy* and *The Second Maiden's Tragedy,* with *Hengist* he returned to the "Lucrece" plot a third time to parody the motifs of his earlier plays.

As Middleton received it, the legend of Hengist is a story of a national disaster, caused in part by the British King Vortiger's infatuation with a pagan beauty.[80] Middleton altered it, as Shakespeare altered the Apollonius legend, by introducing an Eve/Mary opposition. He compounded the sexual sins ascribed to Hengist's daughter, Roxena—she becomes a whore as well as a seductress—and invented the figure of Castiza, the chaste wife, whom Vortiger first rapes and then publicly shames and imprisons. Restored to liberty and honor by Aurelius, Castiza becomes a symbol of Britain itself.[81]

In Middleton's sources Hengist's daughter is guilty of ensnaring the king, poisoning his son, and working covertly for the Saxons in Britain. Middleton, however, omits any reference to her national loyalties, and in spite of its obvious dramatic potential, refrains from developing the murder: We simply see Roxena in a dumb show (following 4.2) persuading "two saxons" to kill the boy. Instead, Middleton emphasizes Roxena's dangerously corrupt sexuality. He invents the major fact of her liaison with Horsus, and from her first appearance, in the dumb show following 1.1, defines Roxena by her lust ("Enter *Roxena* seemeing to take her leave of *Hengist* her father, But especeally privately & warily of *Hersus* her lover; she departs weepeing" [7–9]). Her spectacular theatrical ability, prefigured here in her "Cuning greife" (Chor.2.9) and manifest subsequently in her impromptu demonstration of chastity (2.3.249–83) and her eloquent oath of virginity (4.2.254–60) is—like her ambition—a function of her sexual incontinence. According to legend Vortiger first meets Roxena at a banquet in her father's castle, where she is the cupbearer, and as Holinshed—Middleton's primary source—tells the story, Hengist contrives the match to further his own ambition (1: 556). Middleton, however, separates the fateful meeting from the banquet scene, and makes Roxena fully responsible for the seduction of the king: Arriving in Britain before Hengist can send for her, she promptly ensnares Vortiger (2.3.176–70;199–210).

Emphasizing the religious difference between the British and Saxons, Middleton uses Roxena as a symbol of pagan infidelity.[82] Thus the first dumb show opens with the discovery of the goddess Fortune on an altar, "in her hand a golden round full of Lotts" (1–2); Hengist and Horsus draw lots, then "kneele and imbrace each other as parteners in one fortune" (5–6). Immediately after this Roxena enters and says farewell to both. The stage action suggests a parallel between the goddess Fortune—notoriously promiscuous—and Roxena: the two men are "parteners in one fortune" and one woman.

After Hengist and Horsus have fought their first battle in Britain, a Gentleman observes, "these saxons bring a fortune w[i]th em / Staines any Romaine success" (2.3.1–2), and Vortiger muses, "Never Came powre guided w[i]th better starrs / Then these mens ffortitudes, yet th'are misbeleevers / Tis to my reason wondrous" (8–10). Soon after Horsus urges Hengist to send for Roxena, "That starr of Germany," "a faire fortunate maide" (2.3.158–59). The connection between Roxena and Fortune is strengthened when a Gentleman tells Hengist of his daughter's arrival in Britain:

> Her heart Ioy ravishd at yo[u]r late successe
> Being ye early morneing of yo[u]r fortunes
> So prosperously now opening at her Coming
> She takes a Cup of gold and midst ye armye
> Teaching her knee a Current Cheerefulnes
> W[i]ch well became her, dranck a liberall health
> To ye Kinges ioyes and yours, the King in pr[e]sence
> Who with her sight but her behaviour Cheifely
> Or Cheife I know not which, but one, or both,
> But hees soe farr bove my expression Caught
> Twere Art enough for one mans time and portion
> To speake him and misse nothinge. (2.3.199–210)

In her youth and beauty Roxena thus comes like an emblem of Hengist's prosperous fortunes. Her "Cup of gold" recalls the "golden round" held by Fortune in the dumb show, and Vortiger, stricken with an idolatrous passion, finds his fortune in the cup she holds.[83] He now shares with Hengist and Horsus "one fortune" and one woman.[84]

Roxena's role as a symbol of evil becomes explicit in the play's apocalyptic final scene, as fire—an image of "heavens wrath at ye Last day" (5.2.4)—consumes Vortiger's castle.[85] The avenging lords tell Vortiger they would have remained loyal: "If from that pagan woman, thoudst slept free / But when thou fledst from heaven we fled from thee" (5.2.71–72). Appropriately Roxena, the lustful pagan, burns to death, and as she dies Vortiger's vengeful epitaph identifies her as the Whore of Babylon: "[O]h mysticall harlott / Thou hast thy full due, whom Lust Crownd queene before / Flames Crowne her now, ffor a triumphant whore" (199–201).[86] The hellish coronation imagined by Vortiger parodically inverts the heavenly coronation of the Virgin and the virgin martyrs.

Diametrically opposed to Roxena in every respect, Castiza is chaste and passive, manipulated by a succession of men. Her brief appearance in act 1 foreshadows her sexual exploitation for political ends throughout the play: Castiza enters reluctantly (1.2.150), drawn by her betrothed, Vortiger, into a form of prostitution—she must pretend to be the bride chosen for the virginal Constantius. Unlike Roxena, she is incapable of deceit, though nevertheless obedient to Vortiger's commands. (Castiza's single show of strength—her insistence on taking monastic vows in spite of Vortiger's opposition [2.1.1–34]—is inspired by Constantius' praise of virginity [1.2.170–87]). At her next appearance, in the dumb show following 2.1, she is compelled by her family to marry the newly crowned Vortiger. Middleton's directions make the violence of the enforced marriage clear: "[T]hen ye Lordes, all seemeing respect, Crowne *Vortiger,* then bring in *Castiza,* who seems to be brought in unwillingly by *Devon: & Stafford* who Crowne her and then give her to *Vortiger,* she going forth with him, w[i]th a kind of Constraind Consent" (15–20). The juxtaposition of Vortiger's illicit coronation and coercive marriage suggests a political/sexual analogy. Raynulph's choric commentary highlights the complicity of Devonshire and Stafford—Castiza's father and uncle, respectively—in both events:

> Then Crowne they him, and force ye mayd,
> That vowd a virgin life to wedd,
> Such a strength greate power extendes
> It Conquers fathers, kinn, and friendes. (Chorus.3.11–14).

In *Titus Andronicus, Cymbeline,* and *The Second Maiden's Tragedy,* the heroine's father attempts in vain to force her marriage with the tyrant figure; here he actually succeeds.[87]

When we next see Castiza (3.2), she is again the victim of Vortiger's sexual aggression. Middleton structures the scene metatheatrically: We watch Vortiger watching Horsus playing the role of a rapist (Horsus sets up the rape, assuming the persona of an anonymous gallant, and Vortiger takes over to complete the assault). Vortiger's comments—"I admire him" (96), "By this light heele be taken" (108)—call attention to the crime as a piece of bravura theater. Blindfolded and confronted with Horsus's demands for "love," Castiza responds in an exemplary manner, fainting with horror, then kneeling to implore death or disfiguration before dishonor. Like the Lady of *The Second Maiden's Tragedy,* she explicitly believes that to suffer rape is to participate in "an eternall act of death in lust" (81):

> Be Content to take
> Only my sight as ransome for myne honor,
> And where you have but mockd myne eyes with darkeness
> Pluck em out quite, all outward light of bodye
> Ile spare most willingly, But take not from me
> That w[hi]ch must guide me to another world
> And leave me dark for ever, fast w[i]thout
> That Cursed pleasure w[hi]ch would make two soules
> Endure a famine everlastingly. (98–106)

Castiza is eloquent in her anguish, but the metatheatrical structure of the scene works against her. We are directed away from empathy with the victim to admiration for the technical skill of the plotters.

We are also invited to relish the wit of the exploit. As Horsus declares, "[N]ever was poore Ladye / So mockt into false terror: w[i]th what anguish / She lyes with her owne Lord" (117–19). This is the comedy of it: In Jacobean terms, Castiza's terror is "false." The assault is not rape because the assailant is Castiza's husband, and so Castiza, we are to understand, is distraught about nothing. The "rape" is only a delusion, a psychological fiction.[88] Like Iachimo's violation of Imogen, like the Tyrant's violation of the dead Lady, Vortiger's violation of Castiza is a rape-that-is-not-a-rape. In *Cymbeline* the device permits a high degree of eroticism, in *The Second Maiden's Tragedy* the sensational necrophilia, each without the sacrifice of the heroine's chastity. Here it allows the brutal wit of the queen's assault and subsequent humiliation, which Middleton draws out over some 200 lines (4.2.65–269).

Even more than *Cymbeline* or *The Second Maiden's Tragedy, Hengist, King of Kent,* exposes the relentlessly physical definition of chastity in the Jacobean drama and its ideological support for the patriarchal marriage: The chaste woman is one who is vaginally penetrated only by a husband. In the earlier plays this definition operates implicitly. Here it becomes explicit, for the play's tragicomic ending turns upon it.[89] After defeating Vortiger, Aurelius restores Castiza to liberty and clears her name: She is, as he puts it, "firme in honor, neither by Consent / Or act of violence staind, as her grief judges; / Twas her owne Lord abusd her honest feare" (5.2.261–63). Thus, the "violence" of a stranger would have stained Castiza irreparably, but the "abuse" of her husband has not. (Given this moral context and Castiza's religious fears, her response to the revelation of her assailant's identity makes perfect sense. On learning that it was her *husband* who raped her she cries, "it is to greate a ioy for Life" [275].) Instead of staining her, the assault endured by Castiza, like those endured by the virgin martyrs, has increased her value considerably. She is now hailed with "reverence" by Aurelius (268), and when she kneels to him, he kneels to her: "Arise w[i]th me," he declares, "Greate in true ioy and honnor" (270–71). Moreover, like Christ wedding the true church, the Christian Aurelius takes the chaste Castiza as his bride: "[T]o approve thy purenes to posteritie / The fruitfull hopes off a faire peacefull Kingdom / Here will I plant" (279–81).

Aurelius's agricultural image again connects the body of Castiza with the land of Britain. Like Imogen, Castiza has endured the sexual persecution of a political "usurper" and her union with Aurelius, like Imogen's reunion with Posthumus, signifies the kingdom's restoration to health. The triumph of truth is considerably more perfunctory here, however. The weight of the last scene lies with the spectacular internal destruction of the evil triangle (Vortiger, Horsus and Roxena), not with any martial heroism of Aurelius. (Aurelius does not actually do anything except order the fire [5.2.2].) Although Devonshire and Stafford bring in Hengist captive, their own virtue has been radically qualified by their earlier complicity with Vortiger. Castiza herself has none of Imogen's spirit or the Lady's strength. She is the victim of a ruse more cruel than Iachimo's, and subject to an ordeal more humiliating than Imogen's. Mocked by her virtue, she is at best pathetic.

Coda: *Bonduca*

With *Hengist, King of Kent,* the classical paradigm seems exhausted on the Jacobean stage. After Vortiger's rape trick, what more could be done with tyranny and sexual violation? Even earlier, however, the limitations of the Lucrece pattern were signaled by Fletcher's rejection of it in *Bonduca*.[90]

At first glance *Bonduca* seems to bear little relation to the Lucrece legend. Although the play dramatizes a rebellion—a British uprising against the Roman yoke—there is no tyrant figure, no sexually threatened heroine. A look at Fletcher's primary source, however, reveals a symmetry between the Boadicean uprising and the classical paradigm of rape and rebellion. According to Holinshed, the Iceni, a British tribe, revolted after their kingdom was "spoiled" by the Roman captains, their queen "beaten by the soldiers," her daughters "ravished," and "the peeres of the realme bereft of their goods." The British were "chiefly mooved to rebellion by the just complaint" of their queen, "declaring how unseemlie she had been used and intreated at the hands of the Romans."[91] The British forces were briefly victorious, inflicting terrible casualties on the Romans. Thus, as Fletcher found the story in Holinshed, it has significant parallels with the Lucrece story. Like Lucrece, the British queen and her daughters embody the state: The violation and beating of the women symbolize the subjection of the whole community—indeed, of all Britain. Their abuse becomes a *casus belli*.

Potentially then Fletcher's *Bonduca* could have been a remake of Heywood's *The Rape of Lucrece*, with pathetically ravished princesses, heroic Britons and tyrannical Romans. Fletcher, however, declined this opportunity, and instead subverted the "Lucrece" pattern latent in his material: His rape victims refuse the role of passive suffering, his male hero refuses the role of avenger and the Romans are represented more favorably than the British.[92] The raped women are vindictive monsters, who, in contrast to the Roman Lucrece, actively pursue revenge for their injuries. While Fletcher might plausibly have made the British queen a heroic and admirable figure, he chose instead to depict Bonduca as a dangerous virago, a harridan.[93] Like Tullia in *The Rape of Lucrece*, Bonduca unnaturally usurps masculine authority. Her celebration of victory in the play's opening scene introduces us to a world turned upside down: "[A] woman, / A woman beat 'em . . . ," she crows, "a weak woman, / A woman beat these Romanes" (1.1.15–17). However, her final incompetence in the masculine sphere of war dooms the rebellion to failure and the British royal succession to extinction.[94] Like Rowley's Jacinta, Bonduca pursues vengeance with tragic consequences for her nation.

Fletcher's sources say little about the Queen's daughters, beyond the fact of their rape by the Romans. However, Holinshed describes them mounted with Bonduca in a chariot (1: 500), an image that links them with their mother's aggression. Fletcher picks up and magnifies this hint. As he represents them Bonduca's daughters are perhaps worse than she is: They are sadistic fiends. He effectively limits any sympathy for them by choosing not to dramatize the assaults they complain of: he picks up the story after the rapes, at the moment of the Queen's temporary victory over the Romans. We never learn the circumstances of the assaults on her daughters, nor are the assailants individualized: They are simply "Romanes." (The daughters themselves are effectively nameless, defined by their relationship to the queen and their violated bodies: They are, as a Roman puts it, "crackt i'th'ring" [1.2.170–71]).[95] In contrast then to the situation in every other play of the period that involves a rape, the audience here has no sense of the assailants' guilt. The crime remains abstract.[96] What we do see implicitly, however, is the effect of rape on the victims. They are obsessed with exacting revenge—not on certain guilty individuals, but on the Romans in general. Thus they twice attempt to torture captive Romans. Both times they are prevented by their noble uncle Caratach—general of the British forces and hero of the play. Although historically the British leader Caratacus (Fletcher's Caratach) had no place in Boadicea's rebellion, Fletcher makes him central to the play, where he provides a model of masculine honor against which we are to measure the vicious

nature and actions of the three women. Crucially it is the Romans who tes-
tify most eloquently to Caratach's supreme valor; as their general declares,

> [. . .] one single valor,
> The vertues of the valiant *Caratach*
> More doubts me then all *Britain:* he's a Souldier
> So forg'd out, and so temper'd for great fortunes,
> So much man thrust into him, so old in dangers,
> So fortunate in all attempts, that his meer name
> Fights in a thousand men, himself in millions,
> To make him Romane. (1.2.253–60)

The contrast between Caratach's masculine virtue and his nieces' feminine vice emerges clearly in act 3, when the women, using the younger
daughter as sexual bait, ensnare a group of Roman officers. Fletcher shows
the women—armed, like Amazons, with bows and arrows—reveling in their
power over the captive officers, tormenting them verbally. Just as they are
themselves about to execute the prisoners, Caratach intervenes, rebuking his
nieces angrily for their "treachery":

> Those that should gild our Conquest,
> Make up a Battell worthie of our winning,
> Catch'd up by craft? [. . .]
> A woman's wisdome in or triumphs? Out,
> Out ye sluts, ye follies; from our swords
> Filch our revenges basely? Arm again, Gentlemen:
> Soldiers, I charge ye help'em. (3.5.63–69)

The younger one protests, "By ___ Uncle, / We will have vengeance for our
rapes," and Caratach replies, "By ___, / You should have kept your legs close
then" (3.5.68–71). His brutal answer epitomizes with startling frankness a position endorsed by the play as a whole: The rapes were the women's own fault
and they have no right to revenge. When the younger daughter finally pleads
for just "one shot" (with her bow and arrow) at the prisoners, Caratach orders
her "fiddle-string" (i.e., her bow-string) to be cut (81–82). The implied stage
action—a phallic sword severing the taut string—symbolically repeats the rape
in a context that wholly justifies the violation.

One might argue that Caratach condemns the women's *methods* of revenge, rather than the revenge itself; that the terms of his reproach—"From
our swords / Filch our revenges basely?"—implies his own commitment to
revenge. However, Fletcher explicitly distinguishes Caratach's motives for
combat from those of the women during the sacrifice that precedes the climactic battle. It is a visually and aurally impressive scene: The stage direc-

tions call for music, then *"Enter in Solemnity, the Druids singing, the* second Daughter *strewing flowers"* (3.1.1.s.d.), followed by Bonduca, her first daughter, Caratach and others. The Queen and her daughters in turn appeal to the gods for help, each stressing her desire for revenge. Bonduca prays:

> Hear us you great Revengers, and this day
> Take pitie from our swords, doubt from our valours,
> Double the sad remembrance of our wrongs
> In every brest; the vengeance due to those
> Make infinite and endlesse (3–7).

The first daughter specifically invokes the rapes as a cause for vengeance (26–34). The gods, however, give no sign that they hear these petitions and the second daughter weeps out her prayer: "See, heaven, / O see thy showrs stoln from thee; our dishonours / O sister, our dishonors: can ye be gods, / And these sins smother'd?" (50–52). Though *"smoak* [arises] *from the Altar,"* there is still no flame. Caratach then rebukes the women's "fretfull prayers" and "whinings" (53–54), and offers by contrast a fervent prayer specifically addressed to the god of war, not for *revenge,* but rather for a good fight:

> Give us this day good hearts, good enemies,
> Good blowes o' both sides, wounds that fear or flight
> Can claim no share in; steel us both with angers,
> And warlike executions fit thy viewing;
> [. . .] who does best,
> Reward with honor; who despair makes flie,
> Unarme for ever, and brand with infamie;
> Grant this [. . .] 'tis but Justice;
> And my first blow thus on thy holy Altar
> I sacrifice unto thee. (64–76)

The women's prayers advance their grievances as a cause for victory (let right have might); Caratach prays that the best warriors should win honor (let might have glory). Fletcher clearly directs his audience to support the latter: As Caratach strikes the altar, *"A flame arises"* (76.s.d.), signaling divine approval for his prayer and implicit condemnation of the women's vengeful demands. (It cannot be accidental that in this prayer competition the winner, Caratach, echoes the Lord's prayer—"Give us this day good hearts [. . .]".)

The clear contrast in this scene between Caratach's desire for combat, pure and simple—a desire implicitly characterized as masculine, strong and honorable—and the women's desire for revenge—a desire implicitly characterized as feminine, weak and dishonorable—pervades the play. Fletcher establishes it in the opening scene as Bonduca sneers at the enemy—"These

Romane Girls" (1.1.11)—until rebuked by Caratach. When Bonduca protests, "By ___ I think / Ye doat upon these *Romanes,*" Caratach declares,

> Witnesse these wounds, I do; they were fairly given.
> I love an enemy: I was born a souldier;
> And he that in the head on's Troop defies me,
> Bending my manly body with his sword,
> I make a Mistris. Yellow-tressed *Hymen*
> Ne'er ty'd a longing Virgin with more joy,
> Then I am married to that man that wounds me:
> And are not all these *Romane?* Ten struck Battels
> I suckt these honour'd scars from, and all *Romane*. (54–64)

In this homoerotic celebration of masculine combat, the phallic sword is the source of pleasure and the object of desire; the act of wounding, the sexual initiation of a joyful marriage; and the bond between enemy soldiers in battle is a sacred union. As Sandra Clark observes, "in a world where wounds are identified as embraces the rapes of Bonduca's daughters win no sympathy from Caratach" (86). Indeed. Not only were the women's rapes their own fault (they "should have kept their legs close") but according to the logic inherent in Caratach's longing for penetration by Roman swords, the women no doubt desired their violation; should, in fact, be honored by it (the rapists were, after all, Romans). But of course the logic of Caratach's metaphor breaks down at the point of gender difference: What is honorable for men is not honorable for women. As Caratach expresses it, every scar on his body marks a "marriage" made in one of "ten struck battels"—"all Romane" and all "honour'd." There is no problem with promiscuity in this homosocial marriage pattern: The soldier can "marry" again and again, gaining honor with each scar; the more "marriages" the better. He is a hero, not a whore. This is obviously not true for the woman who—like Bonduca's daughters—is sexually wounded by the enemy soldier in a literally erotic encounter. She is dishonored and defaced by the wound. Radically disfigured, she is a whore, not a hero. Thus "masculine" wounds (to the male body, with a sword) are a source of glory; "feminine" wounds (violation of the vagina) are a source of shame.

Fletcher allows the women some redemption in their public suicides: Rather than yield to the Romans they kill themselves on their castle walls. (Bonduca drinks poison, but her daughters—appropriately—use the phallic sword.) However, as Sandra Clark points out (87–88), the emotional weight of the play lies with the tribute paid to Penyus, the Roman general, who kills himself in the preceding scene, and with the death of the boy Hengo, heir to the British throne. Here, as in *Valentinian,* virtue is understood in martial terms, and the bond between soldiers is the strongest af-fective force. Nationality is less significant than gender: Noble Roman soldiers pay homage to Caratach, and Caratach pays tribute to them. The play ends with the Roman commander embracing him and ordering that "through the Camp in every tongue, / The Vertues of great *Caratach* be sung" (5.3.202–3).[97]

❀ ❀ ❀

Bonduca thus seems to invert the classical paradigm: In contrast to the exemplary Lucrece, the raped women ignobly seek their own revenge; Caratach, their uncle and guardian of the family honor, denounces their cause and literally embraces the enemy. The contrast between this play and the other "Lucrece" plays is more apparent than real, however. Like Heywood's *The Rape of Lucrece,* Webster's *Appius and Virginia,* and Fletcher's own *Valentinian, Bonduca* celebrates male heroism, idealizing the homosocial world of the military camp at the expense of the female victims. The "hug" that Brutus longs to give Tarquin on the battlefield (Heywood 251) is fulfilled in the embrace of Caratach and the Roman general (5.3.190.s.d.). While Brutus is checked in his ardor by the memory of Sextus's crime, however—"Hadst thou not done a deed so execrable / That gods and men abhorre, ide love thee *Sextus*" (251)—Caratach does not even pay lip service to the importance of the assaults suffered by his nieces. His frank denial of rape ("you should have kept your legs close") makes explicit the denial implicit in Heywood's bawdy "catch," *"Did he take faire Lucrece by the toe man?"* (232–33).

Linked to this celebration of masculine valor is a corresponding demonization of the raped woman. In other plays this demonization is more subtle: The victim is represented ambivalently, as both an idealized martyr, and—as a consequence of the rape—in some way monstrous. Lavinia, Lucrece and Lucina are all "sympathetic" victims, but their loss of chastity takes them beyond the pale of their communities. The same patriarchal discourse of female chastity that covertly demonizes these three operates overtly in *Bonduca* where the Queen's daughters are simply monstrous: There is no sympathy for, no idealization of these victims. Like Jacinta, they have failed to accept responsibility for their rapes, failed to accept the culturally prescribed role of self-destructive martyr; even worse, they usurp the masculine role of avenger.

Bonduca thus provides an important perspective on the motif of rape and revenge in the "Lucrece" plays: Read in conjunction with these, *Bonduca* suggests that the victim of sexual assault is sympathetic only to the extent that—like Lucrece, Virginia and Lucina—she is self-destructive; and that the woman who pursues revenge for a rape—like Philomela, Jacinta and the daughters of Bonduca—is demonic, not heroic.

In spite of their wide diversity, all the plays examined in this chapter are united by what I have called the contradictory logic of chastity, an ideology that holds a woman morally accountable for a physical violation over which she has no control. The penalty for this violation is, by convention, death—either immediately or at length, after some form of demonic metamorphosis. Like Lucrece, Lord Antonio's wife and Lucina die swiftly, willingly and appropriately after they lose their chastity. Like Lavinia, Jacinta and Bonduca's daughters survive their rapes only to be demonized before dying. By contrast, Castiza in *Hengist, King of Kent,* who also suffers forcible intercourse, emerges chaste and unpolluted in a happy ending, since her assailant has a legal right to that intercourse. Imogen, a potential Lucrece, survives Iachimo's violation of her chamber, blemished in name but—crucially—not in body. Like Virginia, the Lady of *The Second Maiden's Tragedy* dies undefiled, heroically preserving her husband's sexual property from depredation—at least temporarily. All of these women, except Bonduca's daughters, are celebrated for their exemplary chastity, but the question that determines their fate is not moral but material; not "is she *morally* chaste?" but "is she—or can she remain—*vaginally* chaste?" If the answer to this second question is no, if she is no longer physically pure, or her physical violation seems imminent, then she must die. In the final chapter, I turn to plays that work out a fascinating elaboration of this rule—plays in which the heroine is raped but nevertheless enjoys a happy ending when her assailant acquires a legal right to her sexuality.

Chapter 5 ◉

Redeeming the Rapist

[. . .] your forgiveness will lay me under an eternal obligation to you—
Forgive me then, my dearest life, my earthly good, the visible anchor of my
future hope! [. . .] for in YOU, madam, in YOUR *forgiveness,* are centred my
hopes as to *both worlds:* since to be reprobated finally by *you* will leave me
without expectation of mercy from *above!*

—Lovelace to Clarissa; Samuel Richardson, *Clarissa*

I
n most of the Jacobean "Lucrece" plays a woman's rape or threatened
rape has significant political consequences: Heroic men rally to avenge
the injury and pursue the tyrant-rapist in spite of his rank and the inevitable civic upheaval. In *Valentinian,* however, Fletcher presents the revenge of the wronged husband as culpable and selfish. In *All's Lost By Lust,*
Rowley goes further in challenging the classical pattern of rape and heroic
revenge: Jacinta's malevolent and irresponsible demand for revenge destroys
her country. In *Bonduca,* the Queen's daughters are explicitly blamed for
their assaults, and their desire for revenge is represented as demonic: Their
rapes are, in Baines's terms, wholly "effaced." In my final chapter I turn to
plays that push the logic of "effacement" to a comic rather than a tragic conclusion: *The Queen of Corinth* (c. 1616), by Fletcher, Massinger and Field,
and *The Spanish Gypsy* (1623), attributed to Middleton and Rowley. In both
tragicomedies the disruptive consequences of the crime and its punishment
are contained by a marriage.

Although Gossett terms the marriage of rapist and victim a "shocking" innovation to "the standard rape plot" (309) at a time when laws against rape
were severe, the device was neither truly innovative nor, I suspect, shocking.

Not only, as I contend below, were there precedents on the Elizabethan and Jacobean stage for such a "tragicomic solution" to rape (Gossett 326), but these marriage plays reflect an ambivalence about rape characteristic of both its treatment in the drama generally and in early modern English culture and society at large. If, in spite of the law, rape is commonly treated as a venial sin—a forceful seduction prompted by passion, a sexual act that both parties enjoy, something the woman could have avoided, the subject for a bawdy joke—then the woman who seeks legal redress will appear malevolent and disruptive, especially if, as in early modern England, this effectively means demanding her assailant's life. As Walker notes in her study of "rape narratives" produced in early modern law courts, "men frequently portrayed the women who accused them as malicious, revenge-seeking harpies who 'plotted, practiced and conspired' with like-minded confederates to bring about the 'utter overthrowe and distruccon' of hapless male victims" (4).[1] Given the deep-rooted, if unacknowledged, reluctance to support the prosecution of rape in Jacobean society,[2] it is not surprising that a small group of plays dramatize the forgiveness of sexual assault rather than its punishment.

I argue that rape is not simply trivialized by this strategy—not merely rendered "inconsequential," as Gossett suggests (326)—rather, it is sanctified as part of a providential design. In both plays the rapist is a young man whose sexual sin represents a fall from grace. He is tricked into confessing and repenting the crime by another man, who exercises a providential care for the youth's regeneration. Here the male heroism, which in the "Lucrece" plays expresses itself in a potentially tragic demand for vengeance, becomes the benevolent plotting of the male "spiritual director," whose guile ensures the comic outcome of the play. The role of the heroine—like that of the Lucrece figures—is largely confined to passive suffering. Instead of a glorious death, however, she endures an ignominious half-life after the rape; and instead of redeeming her community, she redeems her rapist. Thus unlike the legendary Lucretia, who demands her ravisher's life, the heroines of the marriage plays give up the right to revenge and instead accept their penitent assailants as husbands—a resolution that signifies the salvation of the errant youths.[3] In a paradigm that celebrates the triumph of the patriarchal will over youthful male lust and the injured female's desire for revenge, rapist and victim are part of a triangle dominated by an older male. Scrutiny of the rape and marriage plot thus reveals an aspect of what Rebecca Bach calls "the homosocial imaginary" that structured the culture of early modern England (504), an "imaginary" that conceived of all social relations (including rape) "in terms of male-male relationships" (518).

The departure of Fletcher and his collaborators from the conventional pattern of rape, suicide, and revenge was not as surprising as Gossett suggests (109). Since Shakespeare had, in a sense, already solved the problem of sexual violation with mercy and marriage in *Measure for Measure,* I begin with his play and its Elizabethan sources.

Measure for Measure

In Shakespeare's problem comedy a young man falls from grace when, prompted by lust, he uses his position of judicial authority to demand a virgin's sexual submission; the disguised ruler (a father figure), assumes the role of "spiritual director" and arranges the villain's public exposure; after he is accused of being a "virgin-violator" (5.1.43), his victim intercedes to save his life. *Measure for Measure,*[4] however, stands apart from the two later marriage plays since technically no rape occurs: The Duke's substitution of Mariana (the villain's intended bride) for Isabella (his intended victim) means that although Angelo *enacts* a rape, he does not *effect* one. The bed trick simultaneously satisfies the requirements for lawful intercourse[5] and transforms the victim into the loving wife necessary for the villain's redemption.

Measure for Measure is based on the legend of "the monstrous ransom" (Lascelles 7), in which a man sentenced to death has a wife or sister who appeals to the magistrate for mercy. After she yields to the magistrate's demand for sexual intercourse as the price of the prisoner's freedom, he nevertheless orders the prisoner executed. The bereaved woman appeals to the ruler, who sentences the magistrate first to marry the woman and then to die.[6] The story thus involves a rape, in which the coercion is emotional rather than physical, and its exemplary punishment: The magistrate pays for his twofold crime with his life, and in most versions the woman receives the magistrate's goods in compensation for her injury (J. H. Smith 387). However, as Shakespeare encountered it in Cinthio's *Hecatommithi* (1565) and Whetstone's *Promos and Cassandra* (1578), the story celebrates the exemplary *forgiveness* of a rape rather than its punishment.[7] Both of Shakespeare's principal sources depart from tradition in having the magistrate reprieved. When the villain faces death, the heroine—who is now his wife—pleads for his release, and the ruler ultimately spares him for her sake. Both works thus turn on the transformation of the heroine from an aggrieved victim into a forgiving wife who literally redeems from death the man who raped her.

Cinthio highlights the problem of rape and its appropriate punishment by making the violation of a virgin the crime for which the heroine's brother is condemned. In her appeal to the magistrate, Epitia pleads her brother's youth, the "spur of love" that drove him to the crime, and his willingness to marry the girl (379).[8] It would be cruel, she declares, "to punish with death a sin that could be atoned for honorably and religiously by making amends for the offense" (380). As she presents it a marriage would make full reparation for the rape, since "the honor of the violated lady" would remain "safe."

Epitia's endorsement of marriage as a solution to rape is worth pausing over since it is based on assumptions that are crucial to the Jacobean marriage plays. As the "Lucrece" plays demonstrate, a woman's chastity is a commodity that can be stolen: Her consent, or lack of consent, to sexual intercourse is finally irrelevant. If a man to whom she is not married attains carnal knowledge of her, she is unchaste. As the marriage plays demonstrate, however, she can regain her chastity if she marries that man. (If either victim or assailant is already married this is obviously impossible.) Underlying this principle is the understanding of women as sexual property: In the act of sex a man takes possession of a woman. Marriage retrospectively converts illicit sexual possession—whether consensual or not—into lawful ownership. From the perspective of the rapist, marriage might seem an acceptance of economic and social responsibility for the victim whose sexuality he has possessed illegally and thus a means of atoning "honorably and religiously" for his crime: This is the way Epitia sees it. From another perspective, however—one that is never voiced in Jacobean drama—such a marriage legitimizes the original act of violation and permits its repetition at will.[9]

Ironically Epitia herself suffers from the magistrate, Juriste, the crime she has excused in her brother.[10] When her injury is doubled with the latter's death, she angrily resolves to stab Juriste, and if possible, "to cut off his head, take it to her brother's grave, and consecrate it to his shade" (383). Thus Epitia wants to slay her *ravisher* rather than herself, and the narrative voice calls her anger "just" (383). Significantly, however, she conceives of the vengeance as a form of piety to her brother. She decides against the plan, not for moral reasons, but because her action might be misunderstood: "[I]t might easily be assumed that she, as a woman dishonored *and so kindled to any evil,* had done that through anger and scorn rather than because he had failed to keep his word" (383, emphasis added). In other words, she does not want to be seen as a victim of *rape* (a woman "kindled to any evil"), but rather as a victim of bad faith.

As Epitia presents her case to the Emperor no reparation is possible, no punishment adequate. When he declares that Juriste must marry her, Epitia is unwilling, since "she could not believe she would ever get anything from him but crimes and betrayals" (385). However, when the Emperor sentences Juriste to death after the wedding, Epitia displays her remarkable "virtue." She reflects that since she is married to Juriste it would be wrong to consent to his death: "It seemed to her that it could be attributed to craving for revenge and to cruelty rather than desire for justice" (386). The wedding ceremony thus transforms her "just anger" into "cruelty" and a reprehensible "craving for revenge." Her cry for justice is silenced by her marriage, her identity entirely subsumed in her role as spouse. Indeed, she so internalizes the force of her marriage oath that she begs for her husband's life and obtains it. To the Emperor it seems

> a wonderful thing [. . .] that she should thrust into oblivion the deep wrong she had received from Juriste [. . .]. And he thought that so much goodness as he saw in that lady deserved that he by his grace should grant her the life of the man who had been sentenced to death by justice. So calling Juriste [. . .] he said to him: "Epitia's goodness, guilty man, has prevailed on me so much that, whereas your crime deserved to be punished not with one death only but with two, she has moved me to pardon your life. This life, I want you to know, comes from her; and since she is willing to live with you, joined by that bond with which I willed you to be bound, I am willing that you should live with her. (387)

The magnitude of Juriste's crime, Epitia's desire for revenge and her reluctance to marry him all heighten the marvelous nature of this forgiveness. Cinthio raises the specter of the raped woman as avenging fury—meditating murder, clamoring for the death penalty—only to exorcize the demon with the marriage rite. The avenging fury becomes an angel of mercy, bringing a second life to the man who "assassinated" her.

Whetstone based his two-part play on Cinthio's novella but made substantial changes, including the Jailor's rescue of the condemned man by the device of a substitute head. He also altered the character of the heroine. In contrast to the powerful and threatening Epitia, Cassandra is pathetic. Her assumption of guilt for the rape begins even before the event (Part 1; 3.7; 327), and her self-hatred afterward anticipates that of Shakespeare's Lucrece, Heywood's Lucrece and Fletcher's Lucina: "Fayne would I wretch conceale, the spoyle of my virginity, / But O my gilt doth make mee blush, chast virgins here to see: / I monster now, no mayde nor wife, have stoupte to *Promos* lust" (Part 1; 4.3; 329).[11] When Cassandra learns of her brother's apparent death and Promos's treachery she is suicidal rather than homicidal, but decides to seek vengeance from the king before killing herself (Part 1; 4.4; 330). Her theatrical death will serve, she hopes, like that of Lucrece, both as proof of her virtue and a spur to justice (Part 1; 5.6; 338). Like Cinthio, Whetstone thus elides justice and revenge: Cassandra's demand for justice is really a demand for revenge (legitimized by the death of her brother).

In spite of her suicidal intentions, Cassandra survives to marry Promos according to the king's command. Like Epitia, she is immediately transformed by marriage into a model wife. Cassandra's "virtue" exceeds that of Epitia, however, since she castigates herself for ever having sought the life of her husband: "T'was I al[a]s, even only I, that wrought his overthroe. / What shall I doo, to worke amends, for this my haynous deede?" (Part 2; 4.4; 362). In this version of the story the heroine's pleas are by themselves insufficient to save the life of the magistrate, and Promos is led to his execution. On the way he exchanges a tearful farewell with the stricken Cassandra, who begs his forgiveness, and promises to atone for her "fault" with a speedy death (5.6; 366–67).

Although the happy ending depends upon the surrender of Cassandra's brother—secretly freed by the Jailor—Andrugio only gives himself up because he learns of his sister's grief. Recognizing the depth of her devotion to Promos ("Whylst that she lyve, no comforte may remove / Care from her harte, if that hir husband dye" [5.5; 365]), Andrugio sacrifices his liberty for Cassandra's happiness. Finally the king pardons both husband and brother for Cassandra's sake: "[L]et them both, thy vertues rare commende: / In that their woes, with this delyght doth ende" (5.6; 369). When Promos asks Cassandra how he can ever repay her, she answers modestly, "I dyd, but what a Wife, shoulde do for you" (5.6; 369).

Like Cinthio, then, Whetstone celebrates the virtue of the woman who saves the life of her violator. He further idealizes—and weakens—this figure by having her assume the guilt both for the rape and for her subsequent complaint to the king. This sanctification of the heroine is matched by a rehabilitation of the villain. Unlike Cinthio's Juriste—whose reformation we have to assume since we learn that he and Epitia live happily ever after—Promos demonstrates his remorse both in a gallows confession (5.5; 365–66) and in a loving farewell to Cassandra (5.6; 366–67). His penitence is rewarded in the final moments of the play when the King restores him to office in words that strongly suggest that Promos's fall—like Adam's—was fortunate: "If thou be wyse, thy fall maye make thee ryse. / The lost sheepe founde, for joye, the feaste was made" (5.6; 369). The biblical allusion points to both the typological resonance of the paternal King, and Cassandra's sacrificial role in reconciling the erring "son" with the righteous "father."

In *Measure for Measure* Shakespeare continued the process of redeeming the villain that was begun in his sources. While Cinthio reprieves Juriste from the *consequences* of his twofold crime, and Whetstone, through the jailor's use of a substitute head, saves Promos from *effecting* the second offence, Shakespeare saves Angelo from committing both rape and murder. Angelo can thus act out his desire to violate Isabella, compound his crime by ordering Claudio's death, experience the consequent guilt and the shame of discovery, only to find that he has offended as "in a dream" (2.2.4): It did not really happen. It is a moral fantasy, in which the "prevenient grace" of the Duke's plot comes between act and effect.[12]

This scheme depends primarily on Mariana's willingness to substitute for Isabella. Barnardine, the designated substitute for Claudio, "will not consent to die" (4.3.54–55), but his refusal is no impediment to the Duke's plan since Ragozine's head is immediately and conveniently available. (Barnardine is rewarded for his self-assertion with his life.) However, the Duke needs a live female body to substitute for Isabella's, and only Mariana's will do, since her status as Angelo's betrothed makes their sexual intercourse "no sin."[13] Unlike Barnardine, Mariana *does* consent to die—to undergo the symbolic death of the rape victim whose identity she assumes in the darkness of Angelo's garden.[14] (She is rewarded for her self-sacrifice with the status of wife.) Thus although the heroine's surrender of her body fails as usual to ransom the condemned man, in this version of the "monstrous ransom" story it succeeds in saving the villain.

Mariana redeems Angelo from his villainy, and the price she pays for his salvation includes not just the physical sacrifice of her virginity and the humiliation involved in taking Isabella's place. It entails the ordeal of the fifth act, in which, according to the Duke's script, Mariana must publicly confront Angelo, at once confessing their sexual intercourse and claiming him as husband.[15] There is a ritual quality to the phrases with which she declares her identity to Angelo ("This is that face [. . .]. This is the hand [. . .] this is the body [. . .]" [206–212]).[16] Her "unmasking" (at line 205) is a parody of a nuptial unveiling: It suggests not the bride's frank gift of herself to her husband, but rather Mariana's public exposure of herself in all her dependence—social, economic, sexual, emotional and spiritual—on the man who has rejected her. For her pains she is again repudiated and slandered by Angelo (he reasserts that "her reputation was disvalu'd / In levity" [220–21]); scorned by the Duke ("thou pernicious woman [. . .] " [240]); and charged with slander. Finally the Duke's script impels Mariana, once married, to kneel again and plead ineffectually for her husband's life.[17] In an effort to win Isabella's help, Mariana even pledges her own life to save Angelo's: "Lend me your knees, and all my life to come / I'll lend you all my life to do you service" (429–30). If we add to the ledger of Mariana's suffering all she endured before the action of the play—desertion, slander and five years of grief in a situation Isabella judges worse than death (3.1.231–32)—then the price she pays for Angelo's redemption seems considerable.

Unlike Whetstone, Shakespeare leaves us to imagine the reformed villain: We have no gallows lament, no request for forgiveness, no promise of better conduct or even a word of thanks. Nevertheless, Angelo's exemplary response to the discovery of the Duke in the Friar's gown—a shamefaced request for "Immediate sentence [. . .] and sequent death" (5.1.371)—suggests an overwhelming conviction of sin.[18] Similarly his anxious reflections in the fourth act ("This deed unshapes me quite [. . .]" [4.4.18–32]) economically convey guilt and regret, if not full penitence.

Shakespeare was apparently much more interested in his villain's temptation and fall—a process that he extends through over 300 lines and two scenes[19]—than in his reformation. During this time Angelo moves from confidence in his own moral rectitude through self-doubt ("What dost thou, or what art thou, Angelo?" [2.2.173]), to a new knowledge of himself ("Blood thou art blood" [2.4.15]), and finally to deliberate vice ("I have begun, / And now I give my sensual race the rein" [2.4.158–59]).

There is an obvious resemblance between Angelo's attempt to coerce Isabella and the efforts of hagiographic tyrants to coerce virgin martyrs[20]—a resemblance implicit in the sources and heightened here by Isabella's status as a novice. However, Shakespeare rewrites the hagiographic encounter between virgin and persecutor from the other side: It becomes primarily the story of the persecutor's temptation, rather than the virgin's. Indeed, through most of her two long scenes with Angelo Isabella is—on the evidence of the text—unaware of his sexual interest in her.[21] Shakespeare thus refuses the opportunity for an explicitly sexual conflict: Isabella is never put to the "virginal fencing" that Marina practices at such length in *Pericles*. He chooses, that is, to dramatize the progress of Angelo's lust rather than the strength of Isabella's resistance to it.

Although Isabella is unaware of Angelo's erotic interest, the audience is not. We watch their first interview through the frame provided by Lucio and the Provost: Their response to Isabella conditions ours. McLuskie comments astutely on the structure of this scene:

> A woman pleading with a man introduces an element of sexual conflict which is made explicit in the bawdy innuendo of Lucio's remarks (II.ii.123–4). The passion of the conflict, the sexualizing of the rhetoric, and the engagement of the onstage spectators create a theatrical excitement which is necessary to sustain the narrative: it also produces the kind of audience involvement which makes Angelo's response make sense. Like Angelo we are witnesses to Isabella's performance so that we understand, if we do not morally approve of, his reaction to it. ("Patriarchal Bard" 96)

We are also privy to Angelo's explicit acknowledgment of desire in his "asides" (2.2.142–43, 158), and share his frightened self-questioning in the soliloquy that closes the scene (162–87).

As Angelo moves from assurance to panic, Isabella—prompted by Lucio as Vice—moves from a meek and decorous hesitation, appropriate to her sex, age, and vocation, to a bold and passionate confidence.[22] It is clearly the tension between her purity (signified by her religious habit) and her passion that Angelo finds provocative.[23] Given the strong cultural association between chastity and silence,[24] Isabella's "prosperous art"—her rhetorical aggression as she warms to the conflict—would, I think, make her virtue problematic for Jacobean audiences as well as for the deputy.

When Angelo appears again at the beginning of 2.4 he is, as Styan points out, "a changed man" (145): He has accepted "the strong and swelling evil" of his lust for Isabella (6). When she arrives alone the audience knows, though she does not, that Angelo is intensely excited ("Why does my blood thus muster to my heart [...]?" [20–23]), and—given the difference between them in sex and power—we will probably see her as vulnerable. The subsequent dialogue is in this way similar to the bedroom scene in *Cymbeline* (2.2): We know that Isabella, like Imogen, is sexually threatened, but we do not know exactly how. (Will Angelo attempt to seduce her? trap her? assault her? control himself?) Where Iachimo has Imogen at his mercy, however, and the eroticism of his approach to her is explicit, Angelo confronts an alert and articulate Isabella and, as he stalks her rhetorically, the eroticism is implicit until almost the end. Like Iachimo's predatory description of Imogen, Angelo's private confession of lust for Isabella, which precedes their dialogue, acts as a lens through which we perceive the heroine as at once innocent and sexually provocative. Because of it we hear the erotic connotations in Isabella's language—"pleasure" (31), "whips" (101), "strip" (102), "longing" (103)—of which she is unaware.

Isabella's response to her dilemma ("Then, Isabel live chaste, and brother, die" [183]) is, of course, technically correct. Any heroine would have to respond in this way when her chastity was threatened, just as the source figures do initially. (Isabella's celebrated "More than our brother is our chastity" is a more strident version of Epitia's "My brother's life is very dear to me, but even dearer to me is my honor [...]" [381].) It is in her dialogue with Claudio that Isabella deviates sharply from the model of the source figures and loses so much of the audience's sympathy. Where Epitia and Cassandra give in to their brother's pleas, Isabella reacts with a virulent anger that most recent critics find pathological.[25] Isabella's response is not sufficiently explained by Shakespeare's desire to preserve the chastity of his heroine: He could easily have made her refusal a sympathetic one, but he chose not to. Isabella's ungracious denial highlights her failure to display the heroic self-sacrifice of the source figures. As Jardine points out, Isabella disturbs us by refusing to conform to our expectations of female heroism: "Were Isabella Lucretia [. . .] she would submit to enforced sex, tell all afterwards, and kill herself. That is what the patriarchy expects of a female hero under such circumstances. [. . .] Shakespeare's Isabella is belittled by the stereotypes to whom she so flagrantly refuses to match up" (191).

Measure for Measure thus demonstrates a significant limitation to the value of chastity: Rape is a fate worse than death for a woman, but it is not worse than her *brother's* death. The explanation for this apparent contradiction lies, I believe, in the implicit assumption of female sexuality as male property. The ethic of chastity encourages women to defend this property even at the cost of their lives, but to surrender it when—as in the case of the "monstrous ransom"—the interests of the male "owner" demand it.[26] As a male Claudio has inherited their father's proprietary interest in Isabella and this accounts for the tacit assumption that she ought to sacrifice her virginity for him. One can test this hypothesis by trying to imagine Claudio as Claudia, a sister rather than a brother: The dramatic tension would be lost. No sister's life is worth the "monstrous ransom": Every brother's is.

Isabella and Mariana both lack the stature of the source heroines, whose role they share. Shakespeare's Duke, however, assumes a much larger role than either Cinthio's emperor or Whetstone's king. After 3.1, he dominates the action.[27] It is he, rather than the women, who achieves Angelo's redemption, though he does it at their expense. One need not read the play as a religious allegory to acknowledge the typological significance of the Duke: He is, for all his manifestly human frailties, a providential figure.[28] He is also, as Daniel Massey points out, the triumphant hero of a melodrama.[29] The fifth act is a theatrical demonstration of the Duke's power over everyone, especially Angelo, Isabella and Mariana.[30] At the end of the play, the glory clearly goes not to the heroine(s), as in the sources, but to the Duke—to whom all hearts are open, all desires known and from whom no secrets are hid.

The Queen of Corinth

The Queen of Corinth[31] owes much to Shakespeare's tragi-comedy—in particular the trick by which a young man is allowed to enact a rape without effecting one. Here, as in *Measure for Measure,* the villain's intended bride substitutes for his intended victim, and so ensures his salvation. Here too a second man assumes the role of "spiritual director" to expose the rapist's crimes, threaten him with justice and then theatrically allow his reprieve. The later play is considerably more sensational than Shakespeare's, however. There is not one but two sexual assaults, both brutal, both like a gang rape: The first is dramatized in a sadistic masque, the second is part of a melodramatic sequence that strongly resembles the climax of a stalker movie. Here, more clearly than in any other Jacobean play, rape emerges starkly as a function of male relationships, female chastity is a function of male property and the victim of rape absorbs the guilt.

In many respects *The Queen of Corinth* conforms to the classical paradigm of the rape plot: In act 1 Theanor, described in the *dramatis personae* as *"a vicious Prince,"* rapes Merione, a *"virtuous Lady,"* and in act 4 her brother and her betrothed lead a rebellion to avenge the crime. Unlike the young Tarquin, however, Theanor does not represent a corrupt government: The eponymous Queen is a *"wise and virtuous Widow"* (10), who shows exemplary justice in sentencing her son to death when, in the last act, he is convicted of rape.[32]

As the play opens we hear three courtiers discussing the latest news: The Corinthian general, Leonidas, has negotiated a peace that depends on the marriage of his sister, Merione, to Agenor, the prince of Argos. The Queen has ratified the treaty for the "common good" even though she had seemed to favor a match between Merione and her own son, Theanor (1.1.38). Moreover, she has forced the prince "with rough threats / To leave his Mistris" and compelled him to "wait upon his Rivall" (1.1.40–43). One courtier asks, "Can it be / The Prince should sit downe with this wrong?" (43–44). His companions reply that *they* would not: "[A] Mother is a name, but put in ballance / With a yong wench 'tis nothing" (45–46). Massinger thus introduces the play's central conflict between mother and son, its sexual focus (the son's right to possess "a yong wench") and the male peers who will support the prince. Theanor also has the support of an evil, older companion called Crates (the name means "force" or "violence" in Greek)—a more sinister Parolles to Theanor's more vicious Bertram.

Tutored by Crates, and aided by the courtiers, Theanor rapes the "wench" on the eve of her wedding to his rival. His crime is a thus a direct transgression against the feminine authority of his mother: It has far more to do with the Queen than with the unfortunate victim, whom Theanor uses to assert his manhood. Unlike Tarquin, then, and most of the tyrant figures in the "Lucrece" plays, Theanor is not driven by lust.[33] The rape is an expression of adolescent male rebellion: In violently exerting power over Merione, Theanor acts out his aggression against the mother who thwarts him.

Theanor's situation at the beginning of the play is similar to that of Bassianus in the first act of *Titus Andronicus:* He is suddenly deprived of his prospective bride, without her consent, for political reasons.[34] Unlike Bassianus, however, who publicly abducts Lavinia to marry her, Theanor simply violates Merione in secret and in disguise. What matters to him is being the first to possess her sexually. He has no interest in her as a wife "to have and to hold": He wants the semiprivate satisfaction of beating his rival to the act of defloration.

The crime itself is dramatized with a degree of macabre sensationalism exceeded perhaps only in *The Second Maiden's Tragedy.* Here, as in the earlier play, the violation of the heroine's body is represented by the violation of a sacred place. Crates selects the temple of Vesta as the site for the "action" (the military and theatrical connotations of the word are both relevant): "The vaults so hollow, and the walls so strong, / As *Dian* there might suffer violence, / And with loud shrikes in vaine call *Jove* to helpe her" (1.1.54–57). At the end of act 1 Merione arrives alone at the temple of Vesta to perform certain prenuptial "devotions" (1.2.96) and as she enters the temple she is seized by four masked men (a conclusion certain to sustain audience interest through the interval).

At the beginning of act 2 Merione enters *"as newly ravished."* Although her initial lament is conventional (she challenges the justice of the gods in terms that recall Heywood's Lucrece), the ensuing action is not. Interrupted by the entrance of Theanor and Crates *"with Vizards,"* Merione appeals to the prince for reparation:

> What ere you are
> Sir, you that have abus'd me, and now most basely
> And sacrilegiously rob'd this faire Temple,
> (I fling all these behinde me) but looke upon me,
> But one kinde loving look, be what ye will,
> So from this hower you will be mine, my Husband. (25–30)

Ironically Merione also appeals to Crates, Theanor's evil genius: "Perswade him like a friend, knock at his Conscience / Till faire Repentance follow" (34–35). Merione's pleas foreshadow the play's "happy" ending (repentance and marriage), in which a reformed Crates plays an instrumental role. The prince's immediate response, however, is to threaten her in dumbshow (he *"Draws his Dagger"* [38.s.d.]). Merione, of course, welcomes death ("come / Ye are mercifull, I thank ye for your medicine" [39–40]). As she continues to invite her murder, six courtiers enter *disguis'd, singing and dancing to a horrid Musick, and sprinkling water on her face* (50.s.d.). She sinks, apparently drugged by the liquid ("my eyes grow dead and heavy" [50]) and with her last breath begs to be spared another rape ("Wrong me no more as ye are men" [51]).[35] When she is "fast" they carry her off.

The scene is clearly calculated to shock the audience: We, like Merione, are surprised by the courtiers' entry, and like her too may apprehend further violence. The explanation for the masque offered by Crates—withheld from us until after the event—highlights the theatrical effect of the action on Merione:

> When she wakes,
> Either what's done will shew a meere dreame to her,
> And carry no more credit: or say she finde it,
> Say she remember all the circumstances,
> Twenty to one the shapes in which they were acted,
> The horrours, and the still affrights we shew'd her,
> Rising in wilder figures to her memory
> Will run her mad, and no man ghesse the reason. (2.3.6–13)

Crates's deliberate use of psychological torture recalls the entertainment sent by Ferdinand to the Duchess of Malfi before her murder: A song sung by a madman *"to a dismal kind of music"* (4.2.60.s.d.) and a dance *"consisting of 8 Madmen, with music answerable thereunto"* (4.2.114.s.d.). Here, as in Webster's play, the masque serves as an ironic wedding entertainment:[36] At the end of the play the rape is retroactively converted to the legitimate consummation of a marriage.

It is, of course, impossible to interpret the masque fully without knowing more about "the shapes," "the horrours," "the still affrights" the courtiers use, what they sing, what gestures accompany the dance. However, we do know that Merione perceives the courtiers as "Furies" (2.3.80; 4.3.79). We may assume then that they are dressed as women, with serpents in their hair or around their bodies, perhaps carrying torches and whips.[37] In classical literature, the Furies attended the queen of the underworld, Proserpina.[38] Merione thus becomes a second Proserpina, ravished to hell, surrounded by her terrifying attendants.[39] This mythological pattern is confirmed by a later description of the crime: The rapist, we hear,

> made day night, and men to furies turn'd,
> Durst not trust silence, vizors, nor her sence
> That suffer'd; but with Charmes and Potions
> Cast her asleep [. . .]
> Acted the Fable of *Proserpines* Rape,
> The place (by all description) like to Hell. (4.3.79–84).

There is, however, more to this scene than Crates's attempt to terrify Merione. The threat of violence implicit in the drawn dagger, the quasi-liturgical effect of the courtiers *"singing and dancing to a horrid Musick"* as they sprinkle water on the victim, and the location of the scene before the temple of Vesta all suggest a ritual sacrifice, a demonic parody of a religious ceremony.[40] The action clearly repeats in symbolic terms the violation Merione has already suffered, and thus allows Fletcher to represent the offstage rape.

The masque also emphasizes the isolation of the heroine *vis-à-vis* the men who overwhelm her twice and underlines the major role of Theanor's accomplices: This resembles a gang rape, though only the prince "enjoys" Merione. As such it is very different from Tarquin's bedroom encounter with Lucrece, or even the semipublic assaults undertaken by tyrants with the help of their panders (*The Second Maiden's Tragedy, Valentinian, All's Lost By Lust*). It is an impersonal sport. (This is indeed Crates's word for the crime: He urges Theanor to be fearless, if he "would have more of this sport" [2.3.23]).[41] The collective nature of the "sport" simultaneously confirms the loyalty of the courtiers to Theanor rather than to the Queen and, since he demonstrates his virility and his independence of his mother, the prince's right to that loyalty. Initially Theanor has misgivings about the rape (he doubts "whether 'twere more manly / To dye not seeking helpe, or that help being / So deadly, to pursue it" [1.1.63–65]), but agrees to the crime after Crates casts him as the mature ruler:

> Now you are a Prince
> Fit to rule others, and in shaking off
> The Bonds in which your Mother fetters you
> Discharge your debt to nature [. . .]. (1.1.77–80)

Like the gang rape practiced by a modern fraternity the ritual affirms the bonds among the males through its exercise of sexual power over the humiliated female.[42]

Although on one level Crates's decision to costume his accomplices as Furies is fully explained by his desire to terrify Merione, the crossdressing this involves is striking. It heightens the masque's transgressive force: In disguising themselves as punishing goddesses the courtiers both appropriate and mock the power of the Queen (the threatening mother). Moreover, since in classical literature the Furies above all avenge parents wronged by a son (Rose 84), Crates's choice of costume for the courtiers is doubly ironic—even witty.

❧ ❧ ❧

After the rape Merione, like Proserpina, is trapped in a kind of nether world. Although not physically dead, she has lost her identity. "I am now I know not what," she cries to the assembled court, "Pray ye look not on me,/ No name is left me [. . .]" (2.3.90–91). Removed from the patriarchal order of widow, wife or maid, she is a pariah. Although Agenor generously pronounces her will unstained and offers to marry her anyway, Merione—like Lucrece—has internalized the ethic of chastity too deeply to allow this:

> [. . .] so unfit and weak a Cabinet
> To keep your love and vertue in am I now,
> That have been forc'd and broken, lost my lustre,
> (I meane this body) so corrupt a Volume
> For you to study goodnesse in, and honour,
> I shall intreat your Grace, confer that happinesse
> Upon a beauty sorrow never saw yet:
> And when this grief shall kill me, as it must do,
> Only remember yet ye had such a Mistris. (157–65)

She announces her intention of living "a poore recluse Nun" till heaven sends her comfort or death (170–72). At this point Theanor could, if he wished to marry her, attempt to claim Merione, with or without confessing to the rape. He has heard Merione's appeal to her rapist, and thus can be sure that a confession coupled with an offer of marriage would entail nothing more serious than his mother's temporary displeasure. Instead, however, Theanor feigns outrage at the crime and allows Merione to retire grief-stricken and ruined.

The physical "corruption" that renders Merione unfit for marriage also renders her unfit for society at large. Indeed she sees herself as a danger, a source of moral infection:

> Let no good thing come neare me, vertue fly me;
> You that have honest noble names despise me,
> For I am nothing now but a maine pestilence
> Able to poison all. Send those unto me
> That have forgot their names, ruin'd their fortunes,
> Despis'd their honours; those that have been Virgins
> Ravish'd and wrong'd, and yet dare live to tell it.
> [. . .] Send those sad people
> That hate the light, and curse society;
> Whose sad thoughts are Graves, and from whose eyes continually
> Their melting soules drop out, send those to me;
> And when their sorrowes are most excellent,
> So full that one grief more cannot be added,
> My Story like a torrent shall devoure 'em. (.3.101–14)

Her self-identification as "pestilence" marks Merione's incipient transformation from Proserpina (innocent victim) to Fury (avenging demon). The image of her story as a devouring "torrent" foreshadows the destructive force of her anger and grief.

When we next see Merione (3.2), she has become one of "those sad people / That hate the light, and curse society." In a house shrouded in black, lit only by "Tragick lights" (3.2.112–114), she nurses her grief. As Agenor observes, "She loathes the light, and men" (60). When her friends—the select few whom she will see—chastise Merione for her "fond" behavior (36), she only replies fretfully "oh" and "pish" (35, 42). We are clearly prompted to criticize her passionate devotion to her wounded chastity. Significantly Agenor demands, "why burns these Tapers now? / Wicked and frantic creatures joy in night" (32–33). Infected by her rape, Merione is indeed becoming a "wicked and frantic creature."

In her transformation from rape victim to "wicked and frantic creature," Merione—like Rowley's Jacinta—recalls the figure of Medusa. As Ovid tells the story (*Metamorphoses* 4.968–79), Neptune raped the beautiful Medusa in Minerva's temple; the angry goddess, blaming the victim, turned her into a Gorgon. Since the Gorgons were closely associated in the Renaissance with Furies (and witches) as types of demonic female,[43] the courtiers disguised as Furies in 2.1 point not only to Proserpina as a model for Merione, but also to Medusa-Gorgon. The connection I am positing between Medusa and Merione is strengthened by the location of the latter's rape in the temple of Vesta. Finally, the presence of the Medusa legend as a "source" for the play emerges explicitly near the end: When the prince, caught in the act of a second rape—again in the temple of Vesta—is brought before the Queen, she asks the silent court "what new Gorgons head / Have you beheld, that you are all turn'd to Statues?" (5.2.48–49).

The destructive potential of Merione's anger—her role as Fury—becomes clear when she, tricked by Crates, mistakenly identifies the hero of the play, Euphanes, as the rapist. (Euphanes is adored by the Queen and people, despised by his brother, Crates, and feared by the jealous Theanor.) Rashly assuming Euphanes is immune from justice, Merione threatens suicide in response, implicitly modeling herself on Lucrece ("ha! I have read / Somewhere I am sure, of such an injury / Done to a Lady: and how she durst dye" [138–40]). Agenor and her brother Leonidas quickly take their cue, resolving upon revenge even at the cost of civil war. As Agenor puts it:

> Like *Corinths* double torrent, you and I
> Will rush upon the Land; nor shall the Queene
> Defend this Villaine in his villainy:
> Lusts violent flames can never be withstood
> Nor quench'd, but with as violent streames of blood. (150–54)

Agenor's image of the avengers as a "double torrent" recalls and extends Merione's earlier figure of her story as a devouring "torrent": Now all Corinth is threatened with "violent streames of blood" released by Merione's anger.

Only Euphanes's supreme virtue prevents Corinth's inundation. With exemplary courage and nobility he surrenders himself—his "Countries sacrifice, / An innocent sacrifice" (4.3.53–54)—to the rebels. Alone and unarmed Euphanes subdues them by the sheer force of his integrity: Agenor and Leonidas weep, kneel and surrender their weapons, inviting him to punish their rebellion with death (124–25). The insurrection concludes with royal pardons, embracing and general jubilation as *"the Souldiers lift up* Euphanes, *and shout"* (4.4.s.d.).[44] Field's celebration of the loyal, aristocratic soldier recalls Fletcher's treatment of Aecius in *Valentinian*. Like the suffering of the noble Aecius, Euphanes's patience upstages that of the Lucrece figure. His offer of "Martyrdome" (4.3.67) saves Corinth, while her irresponsible threat of suicide nearly destroys it.

If Theanor's first rape is a form of compensation for his constrained obedience to his mother, his second is primarily an act of aggression against Euphanes, whom he perceives as a rival for both political power and his mother's favor. Once again, however, neither love nor lust for the victim— Euphanes's betrothed, Beliza—has anything to do with the assault. Angered by the failure of the plot to incriminate Euphanes for Merione's rape and jealous of his rival's ascendancy, Theanor decides to revenge himself by raping Beliza on the eve of her wedding ("I long to have the first touch of her too, / That will a little quiet me" [4.4.18–19]). While Crates planned the first rape, this second assault is the prince's own idea, to which Crates gives only his reluctant assent. The switch in roles marks Theanor's rapid decline into vice, and Crates's rebuke underlines it: "Fie Sir, / You'l be the tyrant to virginity; / To fall but once is manly, to persevere / Beastly, and desperate" (4.4.19–22).[45]

Crates's reluctance to assent to the rape of Beliza seems out of character. This is, after all, the man who encouraged Theanor with the prospect of "more of this sport" after the first rape. However, it prepares us for Crates's imminent repentance—a miracle effected by the long-suffering virtue of Euphanes. When, late in the fourth act, our hero interrupts a duel between his best friend Conon and Crates, he takes his brother's side. In a theatrical display of brotherly love, Euphanes kneels to beg Crates's friendship and offer his own (4.4.69–84). (Leonidas and Agenor, who are watching, exclaim, "A most divine example" [85]). Overcome by Euphanes's generosity,

Crates weeps and repents. Conon reveals that the duel was only a device to bring the brothers together (109–12), and as the five men celebrate the reconciliation tears and blood flow freely. This scene of intense male bonding is the emotional climax of the play.[46]

As a token of repentance Crates promises to tell his brother about Merione's rape and of a "new intent": "Wherein," he says, "your counsell and your active wit / [. . .] will be necessary" (4.4.123–24). Euphanes's exultant reply underlines his role as an agent of providence: "My Prophecy is come, prove my hopes true / *Agenor* shall have right, and you no wrong, / Time now will pluck her daughter from her Cave" (125–27). The audience relaxes in the expectation that Euphanes—like the detective hero of a stalker movie—will prevent the rapist from striking again.

This is not the way Euphanes's "active wit" works, however. Instead, he allows the prince to commit a second rape in order to take him in the act. Forensically this seems unnecessary: Given Crates's evidence, there is no need for the second crime. Massinger could even have gained the melodramatic excitement of the trap (in which he anticipates so many crime films) without actually sacrificing his heroine: It would surely be just as effective to take Theanor in the *attempt*. However, he wanted the sensational and theatrical complications of the double rape, and there was thus no question of sparing the victim. Like the Duke in *Measure for Measure,* however, Euphanes does send a *substitute* victim: Merione *"like Beliza"* (5.2.45.s.d.). And like Angelo, Theanor is allowed to believe himself guilty of rape, only to discover he is after all "innocent"—or in Theanor's case, that he has only raped one woman twice, instead of two women once. Unlike Shakespeare, who admits us to the substitution plot, Massinger tricks his audience: The revelation of the victim's identity is kept from us, as from Theanor, until the final moments of the play when it provides the sudden turn from tragedy to comedy.

Theanor's trial is Euphanes's *coup de théâtre:* It is a play about rape within a play about rape. More visually impressive than the trial of Angelo, it begins with a *"Bar brought in"* followed by "Leonidas *with* Merione (*in white*) Euphanes *with* Beliza (*in black*) Queene, Agenor, Conon, Martiall, *with* Theanor, Crates, Sosicles, Eraton. *Lords, Ladies, [Clerke,] Guard"* (5.4.39.s.d.). The women are color coded to match their emblematic roles in Euphanes's play: Merione (cf. Mariana), who pleads for Mercy/Life, in white; Beliza (cf. Isabella), who pleads for Justice/Death, in black. In a highly formal debate, the women initially match phrase with phrase ("*Beliza.* I was Ravish'd, / And will have Justice. *Merione.* I was Ravish'd too, / I kneele for mercy" [62–4]). The comments of Agenor ("A doubtfull case" [71]) and Leonidas ("Such pretty Lawyers, yet / I never saw nor read of" [71–72]), who form an onstage audience, emphasize the theatrical nature of the contention.

Unlike the trial in *Measure for Measure,* which highlights the bond between the women—Isabella kneels in response to Mariana's pleas—this trial highlights their antagonism: Beliza and Merione abuse each other passionately. It is, of course, all pretense, but the audience does not know this: We are offered as entertainment the spectacle of two women fighting bitterly over their rapist. Beliza turns on Merione (her best friend) and implies that she was not really raped: "For hadst thou suffer'd truly what I have done, / Thou wouldst like me complaine, and call for vengeance" (89–90).[47] In turn Merione addresses Beliza as "Bloody Woman" (116).

In spite of Merione's pleas, the Queen, with Cato-like justice, sentences her son to death. Theanor, however, demonstrates his changed nature by offering to marry Merione (and restore her honor) before he dies for Beliza. He thus devises for himself the traditional sentence of the villain in the "monstrous ransom" story, and his mother applauds his noble proposal: "Thou hast made in this / Part of amends to me, and to the world" (177–78). It is at this moment, when the priest and executioner stand ready, and the Queen wishes for "some miracle" to save the prince, that Crates steps forward. With a theatrical flourish that recalls Paulina's "miraculous" intervention at the end of *The Winter's Tale,* he identifies himself as a "Minister" of the gods: "[S]tand not amaz'd, / To all your comforts I will do this wonder" (185–86).

The first part of the "wonder" is Crates's extenuation of Merione's rape. He explains that he and the others were "easily perswaded" to help Theanor "enjoy" Merione:

> being assur'd
> She was his Wife before the face of Heaven,
> Although some Ceremonious formes were wanting,
> [. . .] but when we perceiv'd
> He purpos'd to abuse our ready service
> In the same kinde, upon the chaste *Beliza,*
> Holding our selves lesse ty'd to him then goodnesse,
> I made discovery of it to my Brother [. . .]. (195–203)

According to Crates the difference between the two assaults is one of property: Theanor had a *right* "to enjoy" Merione since he was her husband "before the face of Heaven," if not the law. The Queen finds this explanation comforting: When Euphanes assures her that it is true, she replies, "I would it were" (205). Once Euphanes picks up the story and explains the substitution of Merione for Beliza, the "wonder" is complete: Theanor is no longer a rapist. Like Angelo he is guilty only of having sexual intercourse with his betrothed before the wedding.

The "miracle" worked by Crates and Euphanes also transforms the trial of the prince, with its bitter argument between the rival "rape-victims," into a wedding masque. As Euphanes declares to the Queen: "[T]hese two Ladies in their feign'd contentions, / To your delight I hope, have serv'd as Maskers / To their owne Nuptialls" (220–22). Like *Measure for Measure, The Queen of Corinth* thus ends with a general procession to the altar. Theanor will marry the twice-raped Merione; Euphanes will marry his (still chaste) Beliza; and Agenor, who lost his bride to Theanor, is suddenly claimed in marriage by the Queen herself. Unlike the strained ending of Shakespeare's play, the mood here is unequivocally festive: Agenor accepts the Queen's offer "with all joy" (223) and everyone applauds.

In *Measure for Measure* the Duke's purpose in allowing Angelo to go through the process of sin, exposure and reprieve may be educative: Critics who see Vincentio as a quasi-divine figure certainly believe it is. There is no doubt about Euphanes's purpose, however. It is confessedly redemptive. As he explains to the court in the final moments of the play:

> I hop'd the imminent danger of the Prince,
> To which his loose unquenched heats had brought him,
> Being pursu'd unto the latest tryall
> Would worke in him compunction, which it has done. (5.4.216–19)

Euphanes thus successfully reclaims the fallen prince, restoring him to his noble nature. Here, as in *Measure for Measure,* the moral stratagem covers its inventor in glory: Euphanes, like the Duke, appears wise, benevolent and forgiving as he enables mercy to season justice. Here too it is the substitute victim who pays for the rapist's redemption and the glory enjoyed by the manipulator. Merione's ordeal is considerably worse than Mariana's, however. Where her namesake goes to a sexual encounter that, we presume, is at least not overtly violent, Merione submits to a repetition of her brutal rape. (It is, moreover, a spectacle necessarily witnessed by Agenor and Leonidas who seize Theanor in "the height of his security" [5.1.8]). Like Mariana, Merione is then required to sue publicly for the villain, but for a villain far worse than Angelo: A man who has raped her with extraordinary cruelty, callously abandoned her to ruin and then attempted to rape her best friend.

The Queen of Corinth thus presents a debased version of the sexual motifs in *Measure for Measure*. In contrast to Shakespeare's challenging study of temptation, sin, mercy and justice, the later play is a crudely sensational thriller.[48] Instead of Angelo's complex and painful descent into villainy, we have Theanor's immediate transformation into adolescent thug. Instead of sexual coercion we have "fraternity" rape, enacted in a masque that epitomizes both the misogyny and the celebration of male bonds that underlie the play as a whole. Finally, Shakespeare's problematic bed trick becomes the gratuitous rape trick, the only purpose of which is to provide the forensic excitements of last act.

The Spanish Gypsy

In *The Spanish Gypsy* (1623)[49] the motif of the redemptive rape reaches its acme. Here sexual assault is used neither for suspense nor eroticism, as it is in *The Queen of Corinth* and, to a lesser extent, in *Measure for Measure*. It is simply the dramatic starting point: A young man abducts (literally, "ravishes") a girl in the opening scene and spends the next four acts dealing with the consequences of his lust. Once again, however, the victim plays a merely passive role in the youth's redemption: His return to grace is achieved primarily by the loving deception of his father, and the drama of his repentance strongly recalls the story of the prodigal son.[50]

The play's main plot is based on one of Cervantes's *Exemplary Novels,* "La Fuerza de la Sangre" ("The Force of Blood"), in which a virtuous aristocratic girl is assaulted one night by a dissolute young nobleman.[51] Rodolfo carries her in a faint to his father's house, where he rapes her in darkness: When he releases her neither one knows the identity of the other. Subsequently the girl, Leocadia, bears his son in secret. Six years later the boy is injured in front of the house of Rodolfo's father, who carries him inside to recover. Providence thus leads Leocadia back to the very room in which she was raped. After she tells her story to Rodolfo's parents, his mother tricks him into marrying Leocadia of his own volition. Only then does he learn that she was the girl he raped. After a jubilant wedding feast, everyone lives happily ever after.

Cervantes's story presents a variation on the romance theme of the family reunion: The "force of blood" is so strong that Rodolfo's father is mysteriously drawn to his unknown grandson when the boy is injured. This is *not* a story of spiritual regeneration: Rodolfo shows no sign of penitence or growth. (He never even apologizes to Leocadia.) The happy ending comes about when, by the grace of God, the patience of Leocadia and the guile of his mother, Rodolfo's sexual desires are directed into a socially responsible contract.

The authors of *The Spanish Gypsy* transform this amoral tale into a moral exemplum. The initial circumstances of the rape are much the same as in the source story: A young man, Roderigo, glimpses a beautiful girl in the street and, with the help of friends, abducts her from her parents (1.1). However, the interaction between rapist and victim after the event is substantially different. When Clara and Roderigo are *"discovered"* in a bedroom at the beginning of 1.3, her first words establish the distinctively homiletic tone of the drama:

> Though the black veil of night hath overclouded
> The world in darkness, yet ere many hours
> The sun will rise again, and then this act
> Of my dishonour will appear before you
> More black than is the canopy that shrouds it. (1–5)

The imagery of dark and light that Clara initiates here pervades the play and foreshadows Roderigo's subsequent spiritual awakening.[52]

As a conventionally virtuous victim of rape, Clara understands herself as "infected" and asks for death after dishonor ("so with your sword / Let out that blood which is infected now / By your soul-staining lust" [11–13]). Like Fletcher's Merione, however, and *unlike* Cervantes's Leocadia, she also appeals to her unknown assailant for marriage ("Are you noble? / I know you then will marry me; say!" [13–14]). Roderigo refuses to speak to her but adds insult to injury by offering her gold (25). While the source heroine just pleads pathetically for anonymity and silence, Clara is aggressive in her despair. As Roderigo leaves the room, she clings to him (21–31)—a stage image that foreshadows his inability to elude the consequences of his crime.

Left alone, Clara kneels to invoke "Revenge" (33) and beg the "lady regent of the air, the moon" to lead her "to some brave vengeance" (36–38). By the light of the heavens she discerns a "goodly" chamber and the garden it overlooks. Her bewildered question, "dwells rape in such a paradise?" (43), suggests the primal nature of the sexual sin Roderigo has just committed.[53] Clara also manages to find a "precious crucifix," which she conceals about her person (50). Her prayers for vengeance, her careful search for clues, the cooperation of the heavens in her efforts, all arouse expectations of a revenge tragedy—expectations that only serve to heighten the ensuing comedy of forgiveness.

We do not learn why Roderigo leaves Clara alone—though obviously he must, if she is to steal the crucifix—but when he returns, he is already a changed man: He asks her name (53–54).[54] The question signals a new willingness to see her as a person. Clara, however, refuses to reveal her identity: "You urge me to a sin / As cruel as your lust [. . .] / Think on the violence of my defame" (54–56). Asking only for an anonymous grave, she renews her appeal for death (65–71). Impressed by the girl's response to her violation, Roderigo apologizes:

> [. . .] since I find
> Such goodness in an unknown frame of virtue,
> Forgive my foul attempt, which I shall grieve for
> So heartily, that could you be yourself
> Eye-witness to my constant vow'd repentance,
> Trust me, you'd pity me. (74–79)

Clara's exemplary deathwish thus initiates the process of Roderigo's reformation. By contrast, Leocadia's virtue only provokes Rodolfo to attempt a second rape (317). With this appeal for Clara's compassion, Roderigo effectively reverses their roles: Although she is his victim, he demands her pity. This tactic is successful: She utters no further reproaches.

In spite of his growing remorse, Roderigo refuses to give Clara his name (and thus a means to revenge or reparation). He does, however, say he *would* pursue her honorably if he had not already raped her:

> Trust me, fair one,
> Were this ill deed undone, this deed of wickedness,
> I would be proud to court your love like him
> Whom my first birth presented to the world. (83–86)

Roderigo's words imply that his "deed of wickedness" has left him unlike himself (Kistners 15) and hint at his need for regeneration in a second birth. They also provide the first of many suggestions that the two are a well-matched couple, destined for union.

Roderigo concludes by offering to do anything to make amends, short of identifying himself. In addition to an oath of secrecy and her release, Clara demands his reformation:

> Live a new man: if e'er you marry—
> O me, my heart's a-breaking!—but if e'er
> You marry, in a constant love to her
> That shall be then your wife, redeem the fault
> Of my undoing. I am lost for ever. (96–100)

Her tone is extraordinary: It is the voice of a dying woman addressing a *lover.* Her "heart's a-breaking" because she imagines her unseen, unknown rapist plighting his troth to another (enviable) woman. By implication Clara sees her rape as a *de facto* marriage: Any legal marriage Roderigo makes will be at her expense. He will be plighting a troth that should be hers. Clara, however, demonstrates her selfless virtue by suppressing her desire for reparation and placing the welfare of her rapist (*de facto* husband) above her own. Acknowledging herself "lost for ever," Clara offers her ruin as a sacrifice to his spiritual growth. In less than 60 lines Clara has moved from her prayer for "some brave vengeance" (38) to self-immolation for the sake of her assailant.

Roderigo leads Clara away closely veiled. As the Kistners point out, the stage image suggests her loss of identity through the rape (18). It also ironically reverses the action of a wedding, where the husband first unveils his bride then leads her to bed.

●●●

In Cervantes's story Rodolfo departs for Italy "with as little thought or concern about what had passed between him and the beautiful Leocadia as though it had never happened" (320). In *The Queen of Corinth,* Theanor is wholly unrepentant until Euphanes arranges his exposure; Angelo regrets Claudio's death rather than Isabella's maidenhead; and none of the tyrant-rapists in the "Lucrece" plays experiences remorse.[55] The authors of *The Spanish Gypsy* thus depart from both narrative source and known dramatic models in their portrait of the remorseful ravisher. Roderigo pretends to leave Madrid, but returns "*disguised as an Italian*" (3.1.s.d.) to look for the woman he wronged, and act 3 opens with his lengthy meditation on sin ("A thousand stings are in me: O, what vild prisons / Make we our bodies to our immortal souls! [. . .] [1–30]).

Clara, meanwhile, is pining ("I have fallen; thoughts with disgraces strive, / And thus I live, and thus I die alive" [2.2.4–5]). Like Fletcher's Merione, she is in a spiritual netherworld. Unlike Merione, however, she has not published her shame. Her doting parents, Pedro and Maria, nurture her and the hope of a "noble" revenge (2.2.19). They also nurture the hopes of her noble suitor, Louis, encouraging him to woo her "hard" (3.2.49). Obviously, they do not see Clara's rape as an obstacle to her marriage with another man. Clara apparently does, however: She cannot acquiesce to Louis, and we are surely to understand her reluctance as a sign of her virtue. By implication, she feels that she is inseparably bound to her unknown rapist-husband.

The crucial discovery of Roderigo's identity comes swiftly: In 3.2 Clara faints near his father's house and is carried into the very room of her assault (an experience that thus partially repeats that of the rape). 3.3 opens with

"CLARA *discovered seated in a chair,* PEDRO *and* MARIA *standing by.*" Grief-stricken, Clara demands to know who owns the house and learns that it belongs to Don Fernando, the "corregidor" of Madrid.[56] When Fernando enters, Clara ascertains that he has a son, then produces the crucifix (which she happens to be carrying) (40). Fernando's response is appropriately melodramatic:

> You drive me to amazement! 'twas my son's,
> A legacy bequeath'd him from his mother
> Upon her deathbed, dear to him as life;
> On earth there cannot be another treasure
> He values at like rate as he does this. (41–45)

Like the "ancestral ring" of Bertram in *All's Well That Ends Well,* the crucifix is Roderigo's most precious possession, the symbol of the honor he has lost in a dishonorable sexual transaction. In assaulting Clara he has betrayed not only the religion that the crucifix represents, but the mother who bequeathed it to him. And just as Helena's possession of Bertram's ring proves both the fact of their sexual intercourse and her status as the young Countess, so too Clara's possession of Roderigo's crucifix proves the truth of her story, and her right to succeed his mother.[57]

The playwright spares Clara a narration of the rape. Instead he has her pull from her bosom an account written "in bloody characters"[58] and bid Fernando read it. As he does so she invokes his impartial judgement of her wrongs:

> [. . .] call back the piety
> Of nature to the goodness of a judge,
> An upright judge, not of a partial father;
> [. . .] consider
> What I have suffer'd, what thou ought'st to do,
> Thine own name, fatherhood, and my dishonour:
> Be just as heaven and fate are, that by miracle
> Have in my weakness wrought a strange discovery:
> Truth copied from my heart is texted there:
> Let now my shame be throughly understood;
> Sins are heard farthest when they cry in blood. (52–65)

Clara's appeal calls attention to Fernando's official position as *corregidor,* a magistrate charged with the administration of justice. Like the Queen of Corinth, Fernando has to pass sentence on his own son.[59] His initial response is reassuring ("This is the trumpet of a soul drown'd deep / In the unfathom'd seas of matchless sorrows" [67–68]). Rather abruptly, however, he leaves in order to "lock fast the door" (69), then returns immediately. The obtrusive and unnecessary exit provides a visual echo of 1.3, where Roderigo responds to Clara's pleas by locking her in, and thus calls attention to the contrast between the two scenes: In 1.3 Clara's request for marriage is denied, in 3.3 it is granted.

The visual echo also points to a parallel between the two scenes. When Roderigo returns to the room in 1.3, the emotional balance shifts, and he begins to plead with Clara. So too, when Fernando returns from locking the door, he becomes the suppliant. He begins by urging Clara to sit (71). This suggests that she has been kneeling before him while he read her story; a posture emblematic of his status as magistrate and hers as suppliant. Now he dramatically reverses their positions:

> [. . .] mark me how I kneel
> Before the high tribunal of your injuries.
> Thou too, too-much wrong'd maid, scorn not my tears,
> For these are tears of rage, not tears of love,—
> Thou father of this too, too-much-wrong'd maid,—
> Thou mother of her counsels and her cares,
> I do not plead for pity to a villain;
> O, let him die as he hath liv'd, dishonourably,
> Basely and cursedly! I plead for pity
> To my till now untainted blood and honor:
> Teach me how I may now be just and cruel,
> For henceforth I am childless. (78–89)

Just as in 1.3 Roderigo had rhetorically changed places with Clara, insisting that his future suffering would deserve her pity, so here Fernando rhetorically (and physically) upstages Clara with his demonstration of grief. Clara's protest ("Pray, sir, rise; / You wrong your place and age" [89–90]) emphasizes the startling nature of the role reversal: Patriarchs do not kneel to young girls. Fernando, however, continues his theatrical self-immolation: "Point me my grave / In some obscure by-path, where never memory / Nor mention of my name may be found out" (90–92). His plea for an anonymous death echoes Clara's earlier plea to Roderigo ("I must not leave a mention of my wrongs, / The stain of my unspotted birth, to memory; / Let it lie buried with me in the dust" [1.3.65–67]). The coincidence strengthens the connection between Fernando and Clara that this whole scene is designed to establish: *Both* now appear as victims of the rape, victims whose sufferings have an equal claim to pity.

Clara does not protest this absurdity. Just as in 1.3 Roderigo's plea for pity silenced her reproaches, so here Fernando's grief elicits her exemplary compassion:

> My lord, I can weep with you, nay, weep for ye,
> As you for me; your passions are instructions,
> And prompt my faltering tongue to beg at least
> A noble satisfaction, though not revenge. (93–96)

Because Fernando has offered to surrender his son to justice (death), Clara dares "with faltering tongue" to ask for much less: A "noble satisfaction"—marriage—instead of "revenge." Fernando clutches at her offer—"Speak that again"—and Clara delicately rephrases her request: "Can you procure no balm / To heal a wounded name?" (97–98). Fernando is transported: "O, thou'rt as fair / In mercy as in beauty! wilt thou live, / And I'll be thy physician?" "I'll be yours," Clara answers (98–100). And so they restore each other: Clara by giving up her right to "revenge," Fernando by giving up his right to negotiate a more advantageous match.[60]

This mutual resurrection of Clara and Fernando is part of a larger romance pattern. The *corregidor* had two children: The first was Roderigo, the second a daughter whom he lost years ago in a shipwreck (3.3.35–36). By graciously agreeing to marriage, Clara restores to Fernando both his son, whose life was forfeit to the law, and—symbolically—the daughter who drowned at sea.[61] His immediate and passionate adoption of Clara—"This daughter shall be ours" (102)—strengthens the play's depiction of rape-as-marriage. Roderigo's crime seems increasingly like the providentially ordained means of Clara's union with both father and son.

The scene ends with Fernando's paternal benediction of the recumbent girl:

> Sleep, sleep, young angel,
> My care shall wake about thee. [. . .]
> Night curtains o'er the world; soft dreams rest with thee!
> The best revenge is to reform our crimes,
> Then time crowns sorrows, sorrows sweeten times. (102–7)

By implication Fernando presents Clara's willingness to marry Roderigo as "the best revenge," the reformation of his crime. Leaving her to sleep, he assumes the burden of active "care."

After her apotheosis as sleeping angel, Clara more or less retires from the play. Like Fletcher's Euphanes and Shakespeare's Duke, Fernando takes center stage to reform the villain, and like them too, he uses craft against vice. Instead of confronting Roderigo with his crime, Fernando, for no obvious reason, *tricks* him into marrying Clara. In this the play partially follows its

Spanish source. However, in Cervantes it is Rodolfo's *mother* who, with Leocadia's cooperation, orchestrates the deception: She is a comic Fate, a variation on the "clever woman" of romance tradition. The wedding is the triumph of female manipulation of male desire.[62] There is no suggestion of moral judgement for the original sin of rape.

In *The Spanish Gypsy* the scheme is wholly managed by Roderigo's *father* and is subordinate to a moral framework. The playwright retains the basic device of Cervantes' story, in which the mother shows her son a portrait of an ugly woman, designated as his bride. His aversion to the image prompts him to declare his desire for the beautiful heroine instead. In *The Spanish Gypsy*, however, this deception is introduced through an inset morality play. When he discovers his son in disguise among the gypsies, Fernando commissions him to write a play on a prodigal son theme: An angry father demands that his wastrel son marry, and his son rejects the chosen bride on the basis of her portrait (4.3.8–143). This play within the play does nothing to advance the plot. However, it does allow Fernando to express his righteous anger at his erring son indirectly, and it also allows Roderigo, who takes the part of the son, to act out his rebellion against paternal authority. More importantly, like the "play extempore" between Hal and Falstaff in the tavern (*I Henry IV* 2.4), Fernando's play provides an interpretative key to the larger action. Like the Boar's Head play, it emphasizes the primary importance of the father-son bond and adds a typological resonance to the action. Now the grieving father emerges as the primary victim of Roderigo's sin, and the disruption of their relationship becomes the most serious consequence of the rape.

When the fiction of the inset play dissolves, Fernando springs the trap in earnest. He insists Roderigo must marry the woman of the portrait, and his son kneels on cue to beg instead for "that young pensive piece of beauty" who watched the play, the anonymous Clara (4.3.223). Their wedding, indicated simply by a procession *"from church"* (5.1.s.d.), follows immediately.

At this point the playwright adds another twist to the deception. Instead of revealing the truth about Clara's identity, Fernando tells Roderigo that his wife is "a wanton" (11) and demands to know what sin he has committed to deserve such a fate (12–14). Under the pressure of his father's stern interrogation, Roderigo confesses to the rape, begs forgiveness and finally repents his refusal to marry his victim. As soon as he expresses this regret aloud— "O, had I married her, / I had been then the happiest man alive!" (37–38)— Roderigo finds that it is miraculously assuaged. Clara and her parents emerge *"from behind the arras"* (38.s.d.). Her responsive declaration—"As I the happiest woman, being married" (39)—is both an enigmatic identification of herself as the rape victim and a retrospective declaration of love for her rapist: If only he had married her, she would have been *"the happiest woman [alive]."*

Roderigo does not ask Clara's forgiveness, nor does Clara offer a fresh re-proach. Instead she vows that she will endeavor to "deserve" his love (54–55). Roderigo's vow goes to his parents: "Fathers both, and mother, / I will redeem my fault" (56–57). Fernando, Pedro and Maria bless him in uni-son (57). I think we should imagine Roderigo kneeling here, surrounded by his forgiving wife, father and parents-in-law: A more elaborate version of the prodigal's return, in which the marriage between rapist and victim is wholly subsumed in the (re)union of the larger family. This tableau all but con-cludes the main plot of the play. The entrance of Louis, Clara's old suitor and Roderigo's friend, allows Clara to confirm the providential nature of the marriage: "[H]eaven's great hand," she tells Louis, "that on record / Fore-points the equal union of all hearts, / Long since decreed what this day hath been perfected" (59–62). The rape is now explicitly the divinely ordained means to the union of their equal "hearts."

Like that other, more famous late Jacobean rape play, *The Changeling,* *The Spanish Gypsy* concludes with the reconstitution of the patriarchal fam-ily. As Stockton observes, in *The Changeling* "the sterile, homosocial bond-ing of lawful patrilineal succession is cemented" by Beatrice-Joanna's death; by contrast, in *The Spanish Gypsy,* this "homosocial bonding" is cemented by the heroine's marriage. These divergent paths to the "happy" ending of fam-ily renewal reveal contrasting patriarchal fears about rape. As Deborah Burks argues powerfully, *The Changeling* expresses male anxieties about female de-sire and complicity: What if your daughter/wife pursued her sexual desire, succumbed to sexual coercion but was still able to pass for a virginal bride? Burks concludes:

> Middleton and Rowley's play tries to manage what the law could not when it exposes the falseness of a woman for all to see, but it is ultimately no more successful than the law in allaying the fear that a woman might succeed in de-ceiving her family and friends. It was a fear fed rather than eased by stories like this one. This fear required, but could not be satisfied with, the bloodied bodies of women like Lucrece and Beatrice-Joanna. Its loathing of the vul-nerability of the female body demanded scenes of retribution [. . . .] But the self-condemning, willing death Beatrice-Joanna dies could only increase the anxiety of a culture that set women as the sentinels to guard familial honor. (782–83)

By contrast, the story of Clara, Roderigo and Ferdinand might successfully assuage male fears about the vengeful rape victim: What if you (or your son), prompted by natural desire, forced a virgin? If, like Clara, she is the ideal woman, she will prefer your welfare to her own desire for revenge. She will understand your position entirely; she will pity you; she will gratefully accept an offer of marriage. And, since she is nobly born as well as beautiful, the marriage will be cause for celebration. It will, indeed, be "a match made in heaven." If in "*The Changeling's* nightmare vision, women's desire is deadly and defiling" (Burks 776), in the fantasy world of *The Spanish Gypsy's* main plot, women's desire is wholly subordinate to patriarchal interests. After Roderigo claims her in violence, Clara desires only him, her *de facto* husband. If, as Burks argues, Beatrice-Joanna's death "could only increase the anxiety of a culture that set women as the sentinels to guard familial honor" (783), Clara's marriage works to allay that anxiety. Her vulnerability to rape allows her, through submissive cooperation with his father, to redeem Roderigo, while preserving the honor and advancing the material interests of both their families. If *The Changeling* requires the exorcism of the demonized female from the family circle, *The Spanish Gypsy* assures the audience that there is no such need: Chaste, true and forgiving, Clara is the ideal wife and daughter.

In spite of its melodramatic opening, *The Spanish Gypsy* is not, like *The Queen of Corinth,* a crudely exploitative thriller. Nor is it, like *Measure for Measure,* a problem play. It is a sentimental romance that in its main plot glorifies sexual assault. To the female viewer it offers the comforting fiction that her rape might promote the rapist's redemption, and her suffering have meaning and purpose. To the male viewer it offers the comforting fiction that his victim (or his son's)—if she is worthy of him—will forgive him. To both it offers the prospect of a loving and prosperous marriage through sexual violence.

All three of these marriage plays dramatize the accommodation of sexual assault in a patriarchal society. As Gossett observes, the structure of the rape and marriage plot "identifies rape with all sexual impulse as it is treated in comedy. [. . .] Rather than being a tragic crime rape becomes a comic error by being brought into the social order" (324). I would emphasize, moreover, that while rape is represented as a "comic error" in these plays, it is not represented as a trivial one; on the contrary, it has marvelous, providential consequences. The comedy it precipitates is neither bawdy nor festive, but the "divine comedy" of Christian salvation. Thus in contrast to most of the "Lucrece" plays, where sexual assault is rationalized as the cause of a community's liberation, in these plays it is rationalized as the cause of the assailant's redemption. In each case the crime is mitigated by youth: It is a first fall; and in each case an older man—the benevolent patriarch—engineers the rapist's penitence at the expense of the victim, who literally atones—that is, effects an at-one-ment—for her rape, reconciling the erring youth with his parental judge. Her suffering and her forgiveness in each case allow a reconstitution of the community, based on the renewed submission of the youth to patriarchal authority. In *Measure for Measure* the compliance of Isabella and Mariana bolsters the Duke's prestige and renews his bond with his chastened deputy. In *The Queen of Corinth* the extraordinary compliance of Merione strengthens Euphanes's prestige, regenerates the heir to the throne, and heals the rift between Leonidas, Agenor and the Queen. Finally, in *The Spanish Gypsy* Clara's cooperation with Fernando allows the triumphant reclamation of his son and heir and the legitimate union of two noble families. If, as Diehl argues, the Duke's "representational strategies" (406) effect in the final scene of Shakespeare's play "a sense of community based on shared guilt"(409)—a community of sinners which includes even the Duke—the simpler strategies of Euphanes and Fernando effect a sense of community based on shared joy in the regeneration of the prodigal son.

Cinthio's novella, a prototype for Shakespeare's play, highlights the miraculous transformation of angry victim into redeeming wife. The narrative stresses the justice of Epitia's anger, the difficulty of her choice to forgive and her achievement of the happy ending. Shakespeare's bed trick blurs this transformation, dividing the role of angry victim and forgiving wife. Nevertheless, Isabella makes a deliberate and deeply felt choice for forgiveness. It is Shakespeare's Duke, however, who receives credit for the happy ending. In *The Queen of Corinth* Fletcher, Massinger and Field emphasize the destructive potential of the victim's anger, rather than its justice, and highlight Merione's abasement rather than any ethical choice. There is no scene in which Merione, like Isabella and Epitia, considers the injury she has suffered and decides to forgive her rapist. Euphanes secures her consent to the rape trick offstage, and she reappears only to participate in the mock battle with Beliza for Theanor's life. The credit for saving Corinth and penitent prince goes to Euphanes. In *The Spanish Gypsy,* where the heroine's willingness to forgive the hero *is* celebrated, Roderigo's father is represented as equally a victim of the rape, and forgiveness is a sentimental rather than an ethical or moral issue. Clara never weighs the claims of just anger against mercy. She understands herself as Roderigo's true wife from act 1, and as a true wife prefers his good to her own. Like Euphanes and the Duke, the patriarch Fernando arranges the happy ending.

The evolution of the marriage plot from *The Hecatommithi* to *The Spanish Gypsy* represents a progressive softening of the issues Cinthio raises—in Baines's terms, a progressive "effacement" of rape. Epitia states what must seem obvious to a modern reader: She does not want to marry the man who raped her ("she could not believe she would ever get anything from him but crimes and betrayals" [385]). Mariana, Merione and Clara, however, are all eager to marry the men who have abused them. Mariana's role as substitute victim palliates the problem. But the authors of *The Queen of Corinth* simply evade it: Merione's forgiveness is assumed, not dramatized, and the rape victim is demonized first in Merione's irresponsible demand for revenge on the wrong person, and later in Beliza's impersonation of the merciless accuser. The authors of *The Spanish Gypsy* mystify the issue. Clara's forgiveness springs from her love for her *de facto* husband, and her marriage is represented as the acme of bliss: She is "the happiest woman."

Gossett points out that these plays allow the audience to watch "the fulfillment of a fantasy of rape and yet the guilt attached to the fantasy—and the act—is removed by the final marriage. The fantasy is both permitted and denied, which may account for the appeal of the plot [. . .]" (324). I think she is right, and the appeal she describes is surely gendered, operating more strongly for males. There is another, perhaps less obvious, but more sinister appeal for females, who are invited to enjoy identification with the graciously forgiving victim. That invitation seems particularly strong in *The Spanish Gypsy* where Roderigo's father kneels before Clara, simultaneously empowering her and securing her aid; calls her "as fair / In mercy as in beauty" (3.3.98–99) and promises to look after her: "Sleep, sleep, young angel, / My care shall wake about thee [. . .]" (102–3). What could be more seductive?

At one point in Margaret Atwood's *The Handmaid's Tale,* the narrator, who has endured coerced sexual intercourse repeatedly, tells an imagined male reader: "Please remember: you will never be subjected to the temptation of feeling you must forgive, a man, as a woman. It's difficult to resist, believe me" (144). She concludes,

> Maybe none of this is about control. Maybe it isn't really about who can own whom, who can do what to whom and get away with it, even as far as death. Maybe it isn't about who can sit and who has to kneel or stand or lie down, legs spread open. Maybe it's about who can do what to whom and be forgiven for it. Never tell me it amounts to the same thing. (144–45)

These Jacobean tragicomedies of forgiveness enshrine the gendered paradigm of sexual sin and absolution that Atwood's narrator recognizes but cannot escape. They are finally about "who can do what to whom and be forgiven for it."

Conclusion ❀

The representation of sexual violence in Jacobean drama is both highly conventional and remarkably diverse, embracing tragedy, comedy, melodrama and hagiographic romance, and ranging from the sentimental to the pornographic. In spite of apparent contradictions, it is also remarkably coherent.

What emerges most clearly from a study of sexual assault in these plays is the material nature of chastity: It is a state of physical, not spiritual, purity, and a woman vaginally penetrated by any man but her husband is unchaste. Her consent or lack of consent to sexual intercourse—though theatrically and dramatically significant—is in this sense wholly irrelevant to the question of her "honor." Paradoxically, then, in spite of an ideology that insists on chastity as a woman's chief virtue, this chastity is beyond her moral control: If she loses the *physical* contest with an assailant, she loses her chastity. Such a material definition of chastity clearly derives from and supports the male's traditional right of property in his dependent female kin (whether daughter, sister or wife). Commenting on prosecutions for rape, Clark and Lewis observe that the "main issue to be decided in law is whether a man's actual or potential property rights have been violated by the rapist. [. . .] This is why a man may force sexual intercourse upon his wife without being accused of rape. He is her proper owner and may use her sexuality as he sees fit." (161).[1] As a study of the drama demonstrates, the violation of property rights is also the main issue to be decided in the Jacobean rape plot. The heroine's fate depends on her status as sexual property. Any woman who loses her chastity to a rapist dies, whether swiftly, like Heywood's Lucrece, or after a painful delay, like Rowley's Jacinta. In the exceptions that prove the rule of chastity-as-property, the heroines of *The Queen of Corinth* and *The Spanish Gypsy* find happiness when their assailants retroactively convert illicit intercourse to wedding rites; and Castiza in *Hengist, King of Kent* calls the news that her unknown rapist was in fact her husband "to greate a ioy for Life" [275].

In general sexual assault is represented as on one level a crime of passion, in which the victim's beauty is the instrumental cause: A series of conventional tyrants make lustful demands on their beautiful victims. At

the same time, however, these assaults are often clearly a function of a patriarchal structure in which women serve as objects of exchange between
men: The female body is the medium for a message the assailant sends
other men. Thus in the "Lucrece" plays there is always a political dimension to the assault, as tyrants prey on the sexual property of their male subjects. Often, as in *The Virgin Martyr*, where Sapritius attempts to rape
Dorothea in front of a crowd that includes his son, the assault is overtly
theatrical, designed to display the assailant's dominance to an onstage male
audience. In *The Revenger's Tragedy* the Duchess's youngest son "has played
a rape on Lord Antonio's wife" (1.1.109): The violence is a "sweet sport"
(3.5.77) he practices with his friends, "a throng of pandars" (1.4.43). In
Cymbeline the wager that sends Iachimo to prey on Imogen and commit
his specular rape is purely a function of male rivalry. Likewise, Cloten's
jealousy of Posthumus, and his desire to demonstrate his dominance to the
King, prompt his vow to rape Imogen in his rival's clothes and "knock her
back" to her father's court (3.5.138–47). Similar elements appear in Heywood's *The Rape of Lucrece*, where a wager among the Roman nobles
(whose wife is most fair and virtuous?) precipitates Tarquin's assault; and
in *The Second Maiden's Tragedy*, where the Tyrant is determined to have the
wife as well as the throne of the deposed Govianus. Nowhere, however, is
this aspect of rape more clear than in *The Queen of Corinth*, where both assaults are precipitated by male rivalry and undertaken to enhance the
rapist's status among his male peers.

Defined primarily through her relations with men, the sexually threatened heroine typically lacks a female community: Justina, Lavinia, Lucrece,
the Lady in *The Second Maiden's Tragedy*, Lucina, Jacinta and Castiza in
Hengist are effectively alone among fathers, brothers, husbands, suitors, panders and ravishers. When she does have female kin or friends, these tend—
like Gratiana in *The Revenger's Tragedy* and the stepmothers of Marina and
Imogen—to be dangerous. The friendship of Isabella and Mariana in *Measure for Measure*—parodied by that of Merione and Beliza in *The Queen of
Corinth*—stands as a significant exception to the rule of female isolation.
The bond between the Queen and her ravished daughters in *Bonduca*, like
the bond between Philomela and Procne, presents another kind of exception: They constitute a "monstrous regiment" of destructive women that
contrasts with the idealized male homosocial world of the Roman army.

In three of the earlier plays, all written for the King's Men, increasingly
ingenious assaults permit the playwrights to exploit the tensions and excitements of sexual violation, while maintaining a *physically* inviolate heroine. In
Measure for Measure (1604) the bed trick allows a rape-that-is-not-a-rape:
Angelo believes he is violating a novice nun while he effectively consum-
mates a marriage with his betrothed. More sensationally in *Cymbeline*

(1608–9) Shakespeare stages a symbolic rape: Imogen's unconsciousness during the bedroom scene allows Iachimo's specular violation while preserving her physical innocence. In *The Second Maiden's Tragedy* (1611) the Tyrant's necrophilia takes the logic of this stratagem to its macabre conclusion: The Lady's chaste spirit is unsullied by the Tyrant's lustful possession of her corpse. In a later variation Middleton "wittily" reverses the bed trick of *Measure for Measure*: The rape of Castiza in *Hengist, King of Kent* (1619) is, from a Jacobean perspective, no rape, since her assailant is her husband. In each case the dramatic maneuvers are governed by the same narrowly physical understanding of chastity.

Similarly, these plays are unified by the absolute value given to a woman's sexual purity: Violation is always a fate worse than death, and sexually threatened heroines from Shakespeare's Lavinia (c. 1592) to Miranda in *The Two Noble Ladies* (c. 1622) ask for death before dishonor. The sole instance in which the value of chastity is limited is in *Measure for Measure,* where it seems that although a woman's sexual purity is worth more than *her* life, it is not worth more than her *brother's*: Again, female chastity emerges as a function of male property rights—in this case, the right of a brother to dispose of his sister's sexuality. Shakespeare's problem play is thus the exception that proves the rule of female virtue: In every other play a heroine displays her virtue primarily by resisting sexual assault against all odds; the worse the odds, the greater the virtue.

Several plays dramatize the misogynist fiction, enshrined in Christian hagiography, that the truly chaste woman cannot be violated. Thus the power of Marina's purity miraculously triumphs in spite her subjection to the brothel, and in *The Two Noble Ladies* Justina's holy chastity withstands the devilish power of Cyprian. The highest accolades are awarded to Massinger and Dekker's Dorothea who retains her virginity in spite of repeated torture and at the cost of her life: She is literally sanctified by her ordeal, appearing at the end of the play as a glorious spirit, *"in a white robe, crownes upon her robe, a Crowne upon her head"* (*The Virgin Martyr* 5.2.219.s.d.).[2]

The strength displayed by these latter-day saints is obviously deeply attractive—to us and presumably to their original audiences: We rejoice in Marina's powerful resilience and relish the intransigence of Victoria in *The Martyred Soldier.* However, implicit in this celebration of the "virgin martyr" is a more subtle demonization of the rape victim: If the truly chaste woman can, like Victoria or Dorothea, choose death before dishonor, the raped woman is not truly chaste. Even though in every case the victim is, from a legal perspective, manifestly innocent—outnumbered, surprised, overwhelmed—she nevertheless assumes the guilt for the crime. (As Fletcher's Caratach says to his niece, "You should have kept your legs close" [*Bonduca* 3.5.71].) Thus after her rape Lucina knows herself "lost for ever," worse than her assailant because unlike him she is beyond the reach of grace, beyond prayer (*Valentinian* 3.1.66–67).

This conception of rape as a consequence of the victim's moral weakness is logically separate from, but supports, the conception of rape as morally and physically corrupting. In addition to the fear of pregnancy ("the bastard graff" anticipated by Shakespeare's Lucrece [1061]), there is a generally unspoken expectation of moral collapse: If a woman has allowed herself to be "surprised" once—and no doubt enjoyed the experience—she will have less power to resist the next time. Thus in *The Queen of Corinth* Merione tells Agenor after her rape that she is too "unfit and weak a Cabinet" for him to keep his "love and virtue in" (3.3; 25).

Such perceived moral pollution points to the role of the assaulted woman as scapegoat. Assuming the guilt—collective and individual—carried by the rapist, the victim of rape undergoes a metamorphosis into her monstrous opposite: The demonized witch, scold and whore. She is, as Haber suggests, a "changeling" (7). As the anti-masque of Furies in *The Queen of Corinth* (2.1) implies, Ovid's Medusa—punished for her rape by her transformation into the deadly Gorgon—provided a classical prototype for the ravished heroines of Renaissance drama. Philomela, terrifying in her revenge, provided another. Shakespeare's Lavinia is, like Medusa, physically monstrous— a sight that renders her uncle "a stony image" (3.1.257)—and, like Philomela, morally monstrous as she helps in the grotesque revenge plot. Shakespeare's Lucrece "bears the load of lust" Tarquin leaves behind (734) and under that load becomes demented, consumed by the violence of her desire for revenge. In *The Queen of Corinth* Merione uses the language of infection: After her rape she calls herself "a main pestilence / Able to poison all" (2.3; 23). So too the violated sexuality of Rowley's Jacinta becomes a poison, a "serpentine and deadly aconite" (2.1.130), which destroys her, her rapist and her country. The morally corrupting effect of rape is most clearly seen in *Bonduca* where the Queen's daughters, "crackt i' th' ring" (1.2.271), are malevolent monsters.[3]

Removed from the patriarchal order of "widow, wife, or maid," the raped woman is a pariah. She has no identity but that of whore. As Lucina puts it, "I am now no wife [. . .] / No familie I now can claime, nor Country, / Nor name, but Cesars Whore" (74–77). In the tragic rape plots—Heywood's *The Rape of Lucrece, Valentinian, All's Lost By Lust*—the victim pays for her whoredom, her failure to preserve her socially approved identity, by dying. In the tragicomedies—*The Queen of Corinth* and *The Spanish Gypsy*—she is reintegrated into the social order by marrying her rapist.

This marriage of rapist and victim, engineered in each case by a benevolent older man, is an obvious accommodation of rape within the patriarchal social structure—an accommodation that, I have argued, reflects widespread ambivalence about the prosecution of rape in Jacobean society. In a larger sense, however, most of these plays involve accommodations of sexual assault, attempts to make it socially functional. In three of the "saint's plays" a woman's resistance to assault redeems at least one of the men who threaten her sexually, providing an exemplary model of chastity to an audience on stage as well as in the theater. As the "converted conjurer" Cyprian declares: "[M]y very lust / deserv's a blessed memorie, since that / was the first, though a foule step to this blisse" (1894–95). This kind of rationalization is more problematic in the "Lucrece" plays, however, where the assault is socially disruptive and where in several cases the survival of the victim implicitly presents a dilemma. (In *Appius and Virginia*, *The Revenger's Tragedy*, and *The Second Maiden's Tragedy* the victim's eager self-sacrifice prevents this embarrassment.)

If the sexual pollution of the violated woman makes her, like any whore, an involuntary source of infection,[4] her anger makes her particularly dangerous. Like the spell of the witch, the curse of the rape victim has a peculiar power, as the tyrants in *Appius and Virginia*, *Valentinian* and *All's Lost By Lust* discover. Practically, in these plays that power is linked to the victim's ability to demand vengeance from her male relations. In *Titus Andronicus*, Heywood's *The Rape of Lucrece*, *Appius and Virginia* and *The Second Maiden's Tragedy* the revenge is socially valuable: The heroine's suffering and death redeem her community, and thus the assault appears providentially ordained. The tribute of Virginius over his daughter's corpse insists on the social value of the legendary marytrdoms: "Two [Ladies fair, but] most infortunate, / Have in their ruins rais'd declining *Rome*—/ *Lucretia* and *Virginia*, both renown'd / For chastity" (5.2.192–95). In *The Queen of Corinth* and *The Spanish Gypsy*, where the danger presented by the rape victim is finally contained by her marriage to the rapist, the redemption of the villain serves to justify her suffering, and—once again—the rape itself.

In *Valentinian*, however, the revenge of the wronged husband has only an ambiguous value—the real redemption from tyranny is effected by the suicide of the noble general Aecius—and in *All's Lost By Lust* the revenge of the wronged father is unequivocally destructive: Spain's subjection to the tyranny of the legitimate king is replaced with an even worse servitude to the Moors. In *Bonduca* the victims' uncle Caratach expressly repudiates the obligation of revenge: Undertaken by the women themselves with their mother (and the soldiers they command), their revenge is demoniacal. *All's Lost By Lust* and *Bonduca* are thus remarkable in representing no accommodation of rape: In neither play does it become socially functional, but by the same token, it is in these plays that the rape victim is most clearly demonized. Both tragedies thus make explicit the anxiety about female sexuality implicit throughout these plays.[5] Indeed, the enchanted castle of *All's Lost By Lust*, a symbol of the "faire-fowle" Jacinta, offers a convenient emblem for the ambivalent sexuality of all the assaulted women in the drama. (While due observance of sacred rites would ensure the castle's/virgin's "happy conquest" by the king, "broke up by violence," it brings his destruction.) Powerful for good or evil, her sexuality is potentially hazardous. Where the heroine remains inviolate—as in the "saint's plays"—we see its power for good; where the heroine is "broke up by violence"—as in several of the "Lucrece" plays—we see its destructive power.[6]

I have stressed the similarities between these plays, emphasizing the extent to which they share a common patriarchal discourse of chastity. It should be obvious, however, that I do not regard the plays as equally misogynist, nor equally simple in their endorsement of patriarchal ideology. Shakespeare's plays represent the contradictory logic of chastity in more complex and interesting ways than the others, offering more room for resistance, more choices for the reader and actor. (Isabella is probably the strongest and most articulate of the threatened women, Angelo's descent into lust the most carefully scrutinized.) The Lady of *The Second Maiden's Tragedy* and Miranda of *The Two Noble Ladies* demonstrate a "masculine" heroism that presents an attractive alternative to the passivity of many others; and Rowley allows much sympathy for the spirited Jacinta, though she is punished with a grotesque death. Most disturbing, perhaps, are the plays of Heywood and Fletcher, with their celebration of the male homosocial military world. In *The Rape of Lucrece* this celebration includes the bawdy comedy that denies the reality of the heroine's rape, and in *The Queen of Corinth* it coexists with the particularly repellent rape trick, by which Merione has to endure a second rape to redeem her rapist and regain her chastity.

Since I began this project, the idea that North American society is currently experiencing a male backlash against a perceived female threat has become widely familiar (if not accepted). *Plus ça change*. As I argue in my Introduction, the early seventeenth century also saw a backlash against women, prompted by the various social, religious, political and economic crises of the time—one that expressed itself, in part, in a widespread anxiety about chastity. At such a time the many plays "founded on rapes" or attempted rapes offered diverse audiences diverse pleasures: The solace of fictional worlds where exemplary, beautiful and aristocratic women prefer death to dishonor; where male sexual aggression is a fortunate fall redeemed by the suffering of his victim; and where male heroism, expressed in martial valor or in benevolent craft, restores the patriarchal order. At such a time the sexually assaulted heroines of the Jacobean stage, so variously abused—cursed, beaten, tortured, raped, pursued to death, pursued beyond death—served as desirable scapegoats.

Notes

Introduction

1. Since the 1970s, several feminist critics have addressed aspects of sexual violence in Renaissance literature, and especially in Shakespeare: See especially the articles by Baines, Burks, Dawson, Gossett, Kahn, J. Newman, Stimpson Vickers, and Williams. Detmer's dissertation, as yet unpublished, provides an important addition to the scholarly discussion. Surprisingly, Marsden asserts that attempted rapes are "relatively rare in Renaissance drama as a whole. When rape occurs, as in *Titus Andronicus* or Fletcher's *Valentinian,* playwrights emphasize the need for revenge but do not represent rape as a titillating exhibition. In contrast, such scenes were routine in the drama of the last decades of the seventeenth century, and their function was decidedly erotic or even pornographic" (185). Although Marsden writes astutely about rape in Restoration drama, the contrast she draws here is misleading. "Attempted rapes" are common in Renaissance drama and, as I argue, their function is at times erotic. Pornographic representation of sexual assault does not depend on the "visible femininity" of an actress, though no doubt, as Marsden contends (185), the introduction of female players fostered a proliferation of such scenes on the Restoration stage.
2. See, for example, Digangi's discussion of female sexuality in *Measure for Measure,* where he advocates "reading oppositionally, with the purpose of discovering the kind of female agency" the male terms of the drama would "exclude and restrict" (590). More recently Findlay, attempting to construct a "historicist feminist perspective on the drama" (5), has argued for a reading of Isabella's silence "as an act of resistance rather than consent" (44).
3. "This helpless smoke of words" (1). Detmer makes this claim at greater length in her dissertation, "The Politics of Telling."
4. The term "sexual assault" was introduced into legal discourse by feminist critics of traditional rape law. For discussion see Tong 112–20, and Clark and Lewis 159–70.
5. For discussion of the medieval statutes see Toner 90–94, Bashar 30–32, and studies by Post; Burks usefully addresses questions of complicity and consent in relation to medieval and early modern rape law (765–70).

6. Chaytor argues this (395–96), without, I think, strong evidence. Bashar (41–42) and Wynne-Davies (130–31), who posit a decisive change in the sixteenth century, attach what seems to me an excessive significance to the statute of 1597 withdrawing Benefit of Clergy in the case of rape. Baines, however, suggests that "rape was still very much a crime against property and class hierarchy, *even as* consent became increasingly important. The law's desire to have it both ways—as a crime against property and as a crime against the person—reveals a crisis in the Early Modern construction of woman's subjectivity: She is both property or passive object and a person invested with agency, with the will and discernment that define consent" (72–73).

7. Wynne-Davies argues that after the statute of 1597 "a woman's body in its sexual sense was seen legally to be her own possession and not that of her nearest male relative" (130). Hale's treatise demonstrates the contrary, however. Thus, he declares that if the victim "consent[s] after the rape," she forfeits the right of appeal, but *"her husband, if married, or the next of kin, if single, may have the appeal"* (1: 633; emphasis added). Similarly, according to Hale, if a woman leaves her husband for another man, her husband may charge him with abduction, in spite of her consent (1: 637). Porter notes that "court room practice continued to treat rape as a crime to be settled man-to-man"; and cites a 1745 case of attempted rape, in which the judge ordered the assailant to pay the complainant's husband "for damages 5 guineas, and at the same time [the assailant] entered into a bond penalty [. . .] never to molest John Biggs or his wife any more" (217). See also Erickson 232–33.

8. For a contrasting biological explanation of rape, see Thornhill, Thornhill and Dizinno.

9. Most feminists would agree that sexual assault is neither the result of "natural" male sexual aggression, nor an aspect of an individual's psychopathology; and that it must be understood in the context of socially constructed sex roles. I have found the studies by Clark and Lewis, Sanday, Scully and Marolla, and Vogelman particularly helpful.

10. In addition to Brownmiller and Griffin, Porter cites Mary Daly, Andrea Dworkin and the London Feminist History Group.

11. Porter acknowledges the utter inadequacy of the evidence (221), and then argues from silence: Since neither court records, nor women's diaries, nor the writings of reformers (male and female) indicate otherwise, "we have no reason to think that rape was a particularly prominent act in the pre-industrial world" (222).

12. Dubinsky's *Improper Advances: Rape and Heterosexual Conflict in Ontario, 1880–1929* (1993) in effect admirably meets this demand.

13. Commentators on rape law frequently note this striking discrepancy. See, for example, Bashar 40–41, Clark and Lewis 55–57, Toner 85, De Groot 324, and Anna Clark 47.

14. In Sussex, for example, during the reign of Elizabeth I "over 1,000 cases of larceny were prosecuted, about 150 burglaries, about 100 homicides and only

14 rapes" (33). Since it is unlikely that there were more than seven times as many murders as rapes in Sussex during this period, we may presume that most rapes did not come to trial. Bashar's findings are supported by the work of Cockburn, who found 50 indictments for rape out of 7,544 in sample records for 1559–1625 ("Nature and Incidence" 58); Sharpe, who notes "a virtual absence" of indictments for rape (*Crime* 170); and Bridenbaugh (2 out of 250 convictions for Middlesex in 1624 [367–68]). Only one of the eight women who told the physician Napier that they had been sexually abused by their masters is known to have prosecuted her employer (M. MacDonald 88).

15. In her study of 34 rape accusations from the north of England (1640–1700), Chaytor notes that several of those "who did finally report the rape said that they had hesitated at first because they were afraid of reprisals" (402n9).

16. Ingram, "Ridings" 91. Without knowing more about the case it is impossible to interpret Vizard's joke accurately. However, given the connection between rape prosecutions and marriage explored below in chapter 5, Vizard's mocking transformation of his trial for rape into a wedding is interesting. For instances of the assailant's friends attempting to silence the victim and her family, see the episodes involving Joseph Parkins reported by Underdown (*Fire* 67–68).

17. Of 274 prosecutions for rape between 1558 and 1700, studied by Bashar, only 45 resulted in conviction (31 of the condemned were hanged, 6 were granted benefit of clergy, and 6 were reprieved) (34–35). "There was an overall preponderance of child victims in the cases that came to court, and a tendency to convict men accused of raping children and to acquit when the victim was an adult woman" (40).

18. Bashar points out, however, that juries were quite willing to impose the death penalty for other offences, like burglary. Thus in Kent between 1558 and 1599 150 people were tried for burglary, 88 were hanged and 34 granted clergy (an 80 percent rate of conviction); however, of the 26 men tried for rape, 4 were hanged and 2 granted clergy (40).

19. Medieval studies also show very low rates of appeal for rape: See De Groot 330, Kittel 104, Hanawalt 66 and J. Carter 153. Carter concludes that in the thirteenth century "the legal system itself presented almost insurmountable problems of humiliation for the victim" (153). For the period between 1660 and 1800, Beattie calculates that a rape case "came before the Surrey assizes on average once every year and a half and before the Sussex courts only once every four years"; many of these cases involved attacks on children (126–27). See also Toner 97–98.

20. Similarly, Marsden notes that in Restoration drama "the motivation for rape is blatantly, even crudely sexual, and to emphasize the erotic potential of the rape itself, the rapist's desire is explicitly stated. Coupled with the physical display of the actress, these descriptions [. . .] operate to arouse the audience's desire" (187).

21. The Duke of Florence rapes Bianca in *Women Beware Women* (2.2); D'Amville tries to rape his virginal daughter-in-law, Castabella, in *The Atheist's Tragedy*

(4.3). In the single clear exception to the rule of obvious villainy, the sympathetic young hero of *The Spanish Gypsy* succumbs to an uncontrollable urge in extraordinary circumstances. As feminists have emphasized, actual rapists are often indistinguishable from "normal" men: See Griffin, *Rape* 5; Clark and Lewis 133–36; Lottes 194; Pollitt 29–30.

22. This enduring belief is well documented in modern studies. See, for example, the comment of an American college student (male) quoted by Beneke: "There has to be some point in every rape where the woman relaxes and enjoys it. I'm not saying that ladies *want* to be raped *because* they enjoy it, but there has to be some point where they enjoy it, because it's enjoyable. Sex is enjoyable" (54). See also Scully 105–6, Vogelman 158, Lottes 206 and Brownmiller 346–58. Williams notes the prevalence of this myth in Renaissance drama (102–3); Baines discusses the early modern legal "effacement" of rape through an "equation between conception, sexual pleasure, and consent" (80).

23. See Rubin's classic essay, "The Traffic in Women," in which she revises the work of Lévi-Strauss on kinship. As she observes, "'exchange of women' is shorthand for expressing that the social relations of a kinship system specify that men have certain rights in their female kin, and that women do not have the same rights either to themselves or to their male kin" (177).

24. According to Girard, the desiring subject imitates a model, or "mediator," who may thus also be a rival for the object of desire: "[T]he mediator's prestige is imparted to the object of desire and confers upon it an illusory value. Triangular desire is the desire which transfigures its object" (*Deceit* 17). Several critics have refined Girard's theory, applying it to historically specific constructions of gender and sexuality. See especially Sedgwick 21–27 and Breitenberg 100–1.

25. The Q2 text calls him a "Country Fellow." See Gurr's Introduction to the Revels edition (lxxviii).

26. Since the romance heroine's chastity frequently depends on the sexual restraint of her male rescuer(s), the moment of relief is often also a moment of anxiety for both heroine and reader/spectator: When a "salvage nation" of fauns and satyrs saves Una from the lustful grasp of Sansloy, she "more ama'zd in double dread doth dwell" until they prostrate themselves before her (*Faerie Queene* 1.6.6–12). By contrast, Proteus first saves Florimell from rape, then threatens her with it (*Faerie Queene* 3.8.20–42).

27. As an illustration of the play, the scene is a curious choice, since it foregrounds the minor comic role of the Country Fellow, and shows the play's hero, Philaster, at his worst. Theatrically, however, it is an exciting scene, including the play's only sword fight, and the only moment in which Arethusa's chastity appears, however briefly, to be in danger.

28. In her analysis of Heywood's *A Woman Killed With Kindness*, Bach argues that the play "appealed to that homosocial imaginary which was as unreal to an actual experience of life as the heterosexual imaginary that we live in is unreal to our actual experience [. . .]. But that 'real' homosocial imaginary

structured early modern English life just as our heterosexual imaginary structures our lives and our critical visions" (517–18).

29. In the terms suggested by Girard's analysis of violence, the conflict between the two groups of men allows each group to form a unified whole, expelling contamination in the form of a scapegoat (*Violence* 223–49). Joplin, revising Girard from a feminist perspective, argues that in "the exchange of women" the scapegoat is gendered feminine: "[T]he woman, in exchange, becomes the surrogate victim for the group. Her body represents the body politic" (36). Dinshaw, extending Joplin's critique, contends that "violence between men and the prior masculine identity formation are enabled by an unacknowledged violence against the feminine" (141).

30. Joplin notes that "the denial or erasure of the mother or any articulated community of women is a crucial aspect of the [classical rape] myths [. . .]. Unlike Philomela who has a sister, Lucrece and Virginia have neither mother, sister, nor daughter" (36n20).

31. As Ovid tells the story, Philomela is raped by her bother-in-law, Tereus, King of Thrace, who cuts out her tongue and imprisons her; she weaves her tale and sends it to her sister, Procne; together the women slaughter Procne and Tereus's son and serve him, cooked, to the king (*Metamorphoses* 6.424–674). Jane Newman persuasively argues that the story of Philomela recalls a tradition of female political agency, threatening to patriarchy and thus repressed by Shakespeare in his *Lucrece*.

32. As Joplin notes, "within the Greek tradition, the myth [of Philomela] was used to teach women the danger of our capacity for revenge" (51). She advocates an alternate reading of the story, however, one that rejects the violent ending and focuses instead on the "voice of the shuttle," celebrating "the woman artist who in recovering her own voice uncovers not only its power, but its potential to transform revenge (violence) into resistance (peace)" (53).

33. See *Metamorphoses* 4.968–79 for Medusa's rape by Neptune and her consequent transformation by Minerva into a Gorgon, with snaky locks and fatal looks. Nancy Vickers connects the image of Medusa with "the heraldry in Lucrece' face," observing that "the monstrous becomes the other side of the beautiful; the obsessively spoken part—the face—the other side of the obsessively unspoken but violated part—the genitalia [. . .] and fear of the female body is mastered through polarized figurations that can only denigrate or idealize" (220). See also Joplin 49–51, and Richlin, "Reading"165.

34. Stimpson notes that "the fact of having been raped obliterates all of a woman's previous claims to virtue" ("Shakespeare" 61). Williams discusses this loss of virtue in terms of a tension between a "shame" culture and a "guilt" culture (94–96).

35. Gossett compares the treatment of rape in four late Jacobean plays to that in four earlier plays. In the earlier plays, which draw on the classical models, the raped woman either commits suicide or is killed as a direct consequence of her violation. According to Gossett, the four later plays "abandon the

usual story in ways which undermine the force in any verbal strictures against rape" (305); thus the heroines do *not* commit suicide and, in a "tragicomic solution which makes rape ultimately inconsequential" (326), three of them actually marry the rapist. Gossett's early plays are: Shakespeare's *Titus Andronicus, The Revenger's Tragedy* (attributed to Tourneur), Heywood's *The Rape of Lucrece,* and Fletcher's *Valentinian.* The late plays are: *The Queen of Corinth,* by Fletcher, Massinger and Field; Rowley's *All's Lost By Lust;* Middleton's *Women Beware Women;* and *The Spanish Gypsy* (attributed to Middleton and Rowley).

36. Gossett's list of "experimental" rape plots might be expanded to include Middleton's *Hengist, King of Kent* and *The Changeling.* There is, however, no clear split between early "conventional" rape plots and late "experimental" ones. The tragicomic device of "marrying the rapist" was not, as Gossett argues, a "shocking" innovation introduced by the authors of *The Queen of Corinth* (309): It appears much earlier in Whetstone's pious Elizabethan comedy *Promos and Cassandra* (1578), and reached Fletcher, Massinger and Field through Shakespeare's *Measure for Measure* (c. 1604). See chapter 5.

37. Commenting on the "great demographic, economic, political, and cultural changes" in early modern England, Stimpson observes that "change bred anxiety about loss and loss of control. This attached itself to gender and stories about gender [. . .]" (K. Newman xii). She thus supports Jardine's earlier claim that "the strong interest in women shown by Elizabethan and Jacobean drama [. . .] is related to the patriarchy's unexpressed worry about the great social changes which characterise the period" (6). Similarly, Woodbridge observes that "the ritualized antifeminism that crops up in [. . .] the early seventeenth century points to a scapegoating of women in response to cultural anxieties arising from rapid social change" (*Scythe* 126). Breitenberg argues for an analogous connection between social crisis in early modern England and the "anxious masculinity" he traces in literary texts (17–27). Hutson makes a much more specific link between "the centrality of women to the fiction and drama of the Renaissance" and the function of women as "signs of credit between men" (7) at a time of socioeconomic transition.

38. Clark's study was republished in 1982 with an introductory essay by Miranda Chaytor and Jane Lewis. For other evaluations of her work, see Thirsk 8–15; and H. Smith, *Reason's Disciples* 29, 35n. For general support of her thesis see Wiesner, "Spinning" and Cahn. See also, however, Judith Bennett's essay challenging the paradigm of "a great divide between medieval and early-modern women" (150): She argues that "[i]n the study of women's work . . . we should take as our central question not transformation . . . but continuity. We should ask: why has women's work retained such dismal characteristics over so many centuries?" (165).

39. Maclean 25, Irwin 2–3, Potter 738–39, Davis 65–95.

40. Luther wrote the *Lectures on Genesis* in 1535–36 (*Works* 1: ix)—long after his own marriage in 1525—and they thus represent his mature thought. For a summary of Luther's views on sex and marriage see Lazareth 208–34. For

Calvin's views, see his comments in the influential *Institutes of the Christian Religion* (Bk. 2, c. 8; 1: 348–50), where he declares: "Virginity, I admit, is a virtue not to be despised; but since it is denied to some, and to others granted only for a season, those who are assailed by incontinence, and unable successfully to war against it, should betake themselves to the remedy of marriage, and thus cultivate chastity in the way of their calling" (348).

41. In spite of Calvin's edicts some Puritans did celebrate sexuality within marriage. As Porterfield argues, this celebration involved the depiction of "sexual pleasure in terms of a well-defined, hierarchical relationship between husband and wife that fostered authority and obedience to the social order" (14).

42. See Hill, *Society and Puritanism* (429–66), and Walzer 183–96. While Cahn opposes Stone's "masculinist framework" (201n5, 213n2), her analysis corroborates a "reinforcement of patriarchy" (86–155). More recently Crawford has expressed cautious support for Stone's position (*Women and Religion* 40). See also K. Thomas, "Women" 42–44; Anderson and Zinsser 256–63; Davis 88–91 and case studies by Slater, Friedman and B. Harris.

43. See Houlbrooke, *English Family* 30–35, 118–19; and MacFarlane, *Love and Marriage* 148–73. Both MacFarlane and Houlbrooke argue for the basic continuity of marriage and the family from the Middle Ages through the early modern period, minimizing the significance of historical change, and the effect of patriarchy within the family.

44. Ozment's *When Fathers Ruled* (1986) celebrates the Protestant and patriarchal family. Ezell, in *The Patriarch's Wife* (1987), argues for a wide gap between patriarchal theory and practice, which allowed the wife to wield considerable power on a private level (163). Alison Wall takes a similarly optimistic position in her case study of the Thynne family.

45. Todd (97–117) also finds continuity, rather than change, in marriage doctrine of the sixteenth century. However, she argues for a Protestant debt to humanism in this respect. See also Collinson 90–92.

46. See also Rogers 140–41, and Greaves 252–57.

47. Eisenstein observes:

It was not a new message but a new medium that changed the character of domestic life most profoundly. [. . .] Elizabethans who purchased domestic guides and marriage manuals were not being given new advice. But they were receiving old advice in a new way and in a format that made it more difficult to evade. [. . .] Type-casting in printers' workshops thus contributed to new kinds of role-playing at home. A "middle class" morality which harked back to Xenephon and the Bible was fixed in a seemingly permanent mold. (429)

48. Louis B. Wright (201–27) and Davies survey the prescriptive literature. See also Erickson 232.

49. Schochet, "Patriarchalism" 424. See also Sommerville on the absolutists' use of the family analogy (22–31).

50. Ingram ("Scolding Women") challenges Underdown's figures for the prosecution of scolds and offers another interpretation of the evidence, but does not, I think, vitiate his argument.

51. There is no way of knowing how common charivaris were. Ingram notes that they seem to have enjoyed the unofficial tolerance of gentry. Since there was no serious attempt to suppress them, there is no reason for their appearance in court records ("Ridings" 110–11).

52. In her Scottish study Larner makes the connection between scolding and prosecution for witchcraft emphatically: "No cursing: no malefice; no witch" (97). She concludes that while the "witch may have been socially and economically in a dependent position [. . .] the factor which often precipitated accusations was the refusal to bring to this situation the deference and subservience which was deemed appropriate to the role" (97–98).

53. Keith Thomas links the persecution of witches to the conflict between traditional neighborliness and a growing individualism (*Religion* 662–78). He believes, however, that "[t]he idea that witch-prosecutions reflected a war between the sexes must be discounted, not least because the victims and witnesses were themselves as likely to be women as men" (679). MacFarlane agrees (*Witchcraft* 160–61). However, the social and economic explanation Thomas offers in no way precludes the operation of misogyny, nor does the fact that women participated in the prosecutions. See Hester for a feminist response to Thomas and MacFarlane. For analyses that allow for misogyny as well as other social and economic factors, see Larner, *Enemies* 92–102; Karlsen's study of the New England witchcraft prosecutions (especially 222–51); and Robin Briggs.

54. Margaret Miles examines the representation of women in the work of German artist Hans Baldung during the early sixteenth century (127–139) and finds in it "a new and increasingly explicit visual connection [. . .] between Eve, sex, and death" (127).

55. See, for example, a broadside of the 1560s, reproduced by Tessa Watt (249), entitled "[Some f]yne gloves devised for newyeres gyftes to teche yonge peo[ple to] knowe good from evyll." It shows two gloved hands with sins listed on the palm of the left and virtues on the palm of the right. "Lechery and her fruites" head the long list of sins; a skull provides an emblem of "her" effects. Similarly, a "godly ballad" of 1566 declared "by the scriptures the plagues that have insued whoredom" (339).

56. See especially Paster 23–63.

57. Dubrow identifies a group of poems in the 1590s featuring a heroine whose chastity is threatened by a lustful monarch: The women "variously yield to the monarch's power or, alternatively, valiantly preserve their chastity in the face of it. No fewer than seven poems of this type appeared within the brief span of two years [. . .]" (401).

58. See Hull 106–26; L. B. Wright 465–507; Masek 146–50; Woodbridge, *Women* 18–113; and Benson 205–30.

59. Benson's analysis of the popular controversy largely supports Woodbridge's judgement in this respect. She finds that, in contrast to translations of continental works such as Agrippa's *De nobiltate et praecellentia Foemenei sexus* (David Clapham, *A Treatice of the Nobilitie and excellencye of woman kynde* [1542]), "native English works do not challenge the traditional valuation of women on the basis of their sexual purity [. . .]. They are conservative works, and they reinforce traditional stereotypes of ideal conduct and roles for each sex by means of antimasculine satire and sentimental portraits of women" (205–6).

60. Tessa Watt lists the stories of "Constant Susanna," Sampson and Delilah, and David and Bathsheba as among the most popular ballads of the period (117); Susanna was also commonly pictured in wall paintings and clothes (194, 202, 209).

61. Benson concurs: The popular controversy represents, she declares, "a resistance movement, a commitment to a notion of woman that was being threatened, and the enemy *against* whom the authors defend is the independent woman. They defend against her threat by denying that women either have the capacity for independent action or desire it; they celebrate a docile, chaste, conventional ideal" (206).

62. Louis B. Wright considers it to be "a result partly of King James's dislike of women and partly of what seems to have been the actual extravagance and vanity of the sex" [*sic*] (481). For a feminist perspective and a more thorough treatment of the issue, see Woodbridge, *Women* 110–12 and 139–83.

63. Underdown believes, however, that this strain may be related to a "slight enlargement" of women's economic role in wooded-pasture areas ("Taming" 155–56).

64. Stone, *Causes* 67–117, summarizes the material changes that may have engendered the fear of disorder. According to Cressy, "the Reformation, however slow and fitful that may have been in England, had effects that were profoundly traumatic" (*Birth* 476–77). Social tensions quickened around 1600. See Hill, *World* 39–56 for the disturbing social mobility; Cockburn, "Nature" for the widespread fear of crime; and Hurstfield, MacCaffrey, and Neale 59–84 for the political stress. Tawney covers the "financial debacle" of James's reign (esp. 95–98 and 137–42). For the license of the Jacobean court, see Akrigg 227–47.

65. This is not say that women were simply victims of this process. As the extent and nature of their involvement in witchcraft prosecutions and defamation cases makes clear, they were also agents, who generally accepted and used the misogynistic discourse available to them (Sharpe, "Witchcraft"; Gowing, "Language").

66. For a valuable discussion of scapegoating theory and its application to Shakespeare, see Woodbridge, *Scythe* 86–151. See also Stockton's analysis of Beatrice-Joanna as scapegoat in *The Changeling*. The roles played by men in the plays I examine suggest a connection with the shift Orlin notes in domestic tragedy between 1603 and 1605: "[A] shift to male violations of the

oeconomic order or, at the least, an enlargement of the repertoire of criminal misconduct to include male as well as female" (238). Theatrical appetite,
she argues,

> brought into the foreground the recurrent subtext of the husband's
> responsibility for disarray in his household, figuring him as the new
> locus of disorder. Thus *Othello* and *A Yorkshire Tragedy* represent one
> climax in an extended dialogue between the public stage and popu-
> lar prescription. Their focus on the male transgressor served to re-
> lease the most direct engagement yet of hegemonic ideals of
> household function, and it underlined the dysfunction of those
> ideals. (239)

Similarly, in plays founded on sexual violence males threaten the patriarchal
family in their complementary roles as lustful predators and as inadequate
guardians of female property.

67. As Dolan points out,

> legal and popular representations of infanticide and witchcraft ex-
> plicitly target those women who live outside male supervision, re-
> vealing the anxieties such women provoked. Accounts of infanticide
> and witchcraft demonize women's self-assertions as attacks directly
> on the family they stand outside (and, the logic goes, therefore
> against): Such women slaughter infants, undermine domestic pro-
> duction, and hold secret rites. (14)

68. On the London theater as market, see Bruster, who emphasizes the way in
which market forces "shaped dramatic commodities to answer the various
manifestations of social desire" (3).

69. Although Howard analyzes the anxiety generated by women's attendance at
the public theatres, such women would surely have been in a clear minority.
For what it is worth, only 16—roughly ten percent—of the 162 playgoers
that Gurr has identified between 1567 and 1642 are female (*Playgoing* xv,
191–204). Lady Anne Halkett's testimony underlines both the social constraints and the financial cost that must have been prohibitive for many
women:

> so scrupulous I was of giving any occasion to speake of mee, as I
> know they did of others, that though I loved well to see plays and
> to walke in the Spring Garden sometimes [. . .], yett I cannot re-
> member 3 times that ever I wentt with any man besides my broth-
> ers; and if I did, my sisters or others better than myselfe was with
> mee. And I was the first that proposed and practised itt for 3 or 4
> of us going together withoutt any man, and every one paying for
> themselves by giving the mony to the footman who waited on us,
> and he gave itt to the play-house, And this I did first upon hearing
> some gentlemen telling what ladys they had waited on to plays, and

how much itt had cost them; upon which I resolved none should say
the same of mee. (*Autobiography*; qtd. Gurr, *Playgoing* 195–96)

70. Using the writing of Renaissance women to "recreate an imagined female audience" for the drama (6), Findlay suggests critical readings of several male-
authored texts. While this seems to me a stimulating and valuable project, I
emphasize the hegemonic power rather than the oppositional potential of
the drama. Actual women no doubt responded to these plays diversely, according to their individual circumstances. However, the primary and powerful invitation issued to the female spectator by the plays I examine seems
to be that of identification with the object of male desire. If such identification brought with it limited forms of empowerment for women, this argues
for the drama's ideological power, for hegemony works by securing the consent of the governed (Gramsci 244): A patriarchy must elicit the consent of
its female subjects.

Chapter 1: The Legends of the Saints

1. On sexuality in the early church see P. Brown, Fox 351–74, Bugge 12–21
and Pagels.
2. See E. Clark 43–46, M. Miles 53–63 and McNamara 153.
3. This is the burden of his treatise, *On the Veiling of Virgins* (*De Virginibus Ve-
landis; Opera* 2: 935–62). See especially chapter 7: "Therefore so dangerous
a face should be covered, [a face] which has thrown scandal as far as
heaven . . ." (*"Debet ergo adumbrari facies tam periculosa, quae usque ad
coelum scandala jaculata est . . ."*) (948); and chapter 16, in which Tertullian
admonishes women directly: "All ages [of men] are endangered by you. Put
on the armour of shame, surround yourself with the wall of modesty, raise a
barrier for your sex which will keep your eyes in and others' out" (*"omnes in
te aetates periclitantur. Indue armaturum pudoris, circumduc vallum verecun-
diae, murum sexui tuo strue, qui nec tuos emittat oculos, nec admittat alienos"*)
(960). See also Tertullian's *The Apparel of Women* and Chrysostom's *On the
Necessity of Guarding Virginity* chapter 1, 209–12.
4. For the limited power enjoyed by women in the early Church and the
misogyny of the Fathers see E. Clark 23–60 and 175–208, Reuther, McNamara, Harvey and P. Brown.
5. Compare, for example, the earliest account of Agape, Irene and companions,
from the fourth century (*Acts of the Christian Martyrs* 280–93), with that
given by Aldhelm in the seventh century (*De Virginitate* [carmen]
2194–278; *Poetic Works* 151–53). In the former chastity is at most a minor
theme, in the latter it is primary. A particular emphasis on female chastity is
already apparent in Eusebius's fourth century *Ecclesiastical History,* however.
His account of Potomiaena's martyrdom is prototypical of later legends, in
which chastity becomes equivalent with religious belief: "Boundless was the
struggle she endured against her lovers in defence of her bodily purity and

chastity (in which she was pre-eminent), for the perfection of her body as well as her soul was in full flower. Boundless too were her sufferings, until at last after tortures terrible and horrifying to describe she was consumed by fire. . . ." (bk. 6, ch. 5; rpt. in *Acts of the Christian Martyrs* 133).

6. This remained true of the Western Church, though not of the Eastern. Throughout the Church, however, the overwhelming majority of saints are male and aristocratic or upper-class (S. Wilson 37–38). Larner observes that in "the peak period for saints (thirteenth to fourteenth centuries) sanctity was sex-related to males in much the same proportion as witchcraft was later to females" (94). For historic examples of matrons as martyrs see the story Perpetua and Felicitas (*Acts of the Christian Martyrs* 106–31): Perpetua was a nursing mother and Felicitas, her slave, gave birth in prison.

7. Examples of this type include Agnes, Agatha, Margaret, Lucy, Barbara and Juliana. Katherine of Alexandria and Eufemia represent a variation: They are prosecuted for their faith before their chastity is tested. The lives of Daria and Cecilia illustrate a second variant: They are wedded but not bedded before dying for their faith.

8. Easton analyzes the erotic violence in the iconography of St. Agatha.

9. *"Candida sed rigido violavit viscera ferro:/ Purpureus cruor extemplo de carne manavit"* (*De Virginitate* [carmen] 1832–33; *Opera* 429).

10. See also Hildegard of Bingen's twelfth-century hymn to St. Ursula, in which the moment of the virgin's death is the climax of her love affair with Christ: "[T]he red blood of the innocent lamb streamed out in her wedding" *("rubicundus sanguis innocentis agni/ in desponsatione sua / effusus est")* (*"O Ecclesia"* 9; Barbara Newman reprints the text in an appendix to her study of Hildegard [281]).

11. *The Lives of Women Saints of our Contrie of England* 89–90. Written by an anonymous Roman Catholic scholar (c. 1610–15), this hagiographic collection remained in manuscript until the nineteenth century.

12. See De Voragine 6: 62–68 for one of the most common accounts. John Wilson, in his *English Martyrologie,* records an alternate version (288–89). Baring-Gould traces the history of the legend and speculates about its origins (12: 535–56). Ursula proved an embarrassment to the modern Roman Catholic Church: She and her company were removed from the list of saints in 1969.

13. Similarly Palladius tells the story of a saintly woman who shuts herself in a tomb for ten years, and explains her self-inflicted punishment thus: "A man was distressed in mind because of me and, lest I should seem to afflict or disparage him, I chose to betake myself alive into the tomb rather than cause a soul, made in the image of God, to stumble" (*The Lausiac History of Palladius,* 5.1; cited by M. Miles 199n22).

14. The Church here follows Augustine. The *locus classicus* is Augustine's meditation in *The City of God* (bk. 1, ch. 16–28) on the ethics of suicide to avoid rape. For the position of the Catholic Church in this century see Vermeersch and Camelot.

15. Suicides include those recorded by Eusebius: The woman of Antioch and her two daughters, who drown themselves to forestall rape. As Eusebius tells the story, the mother plays the part of Ursula: "[She] exhorted both herself and her girls that they ought not to submit to listen to even the least whisper of such a thing, and said that to surrender their souls to the slavery of demons was worse than all kinds of death and every form of destruction" (bk. 8, ch. 12; Loeb 2: 289–91).

    One of these daughters became enshrined in legend as St. Pelagia (Baring-Gould 6: 89). For disfigurement see the legend of Orbila and Modwen (*Lives of Women Saints* 92), and the story of Ebba and her companions (J. Wilson 234). See also Schulenburg on the medieval practice of self-mutilation among religious women.

16. The profound ambivalence of the Church on this issue is rooted in patristic dissent (Schulenburg 32–38). In his *De Virginibus* Ambrose implicitly endorses suicide in defence of chastity (bk. 3, ch. 7; *Opera* 241–44), and Jerome, in his commentary on *Jonah*, explicitly exempts such suicides from the general prohibition of self-murder (*Opera* 1129). Aldhelm cites Jerome enthusiastically in his prose *De Virginitate*, chapter 31 (*Opera* 269).

17. "*Tolerabilius est mentem virginem quam carnem habere*" (*De Virginibus* bk. 2, ch. 4; *Opera* 225).

18. "*Christi virgo prostitui potest, adulterari non potest. Ubicunque virgo Dei est, templum Dei est: nec lupanaria infamant castitatem, sed castitas etiam loci abolet infamiam*" (*De Virginibus* bk. 2, ch. 4; *Opera* 225).

19. This model of female heroism, though strongly endorsed by the Church, was not of course exclusively Christian. Heroines of the Greek romances regularly demonstrate their worth by resisting assault, though they do not usually die as a result. In *Clitophon and Leucippe,* a romance of the late third century by Achilles Tatius, translated into English in 1597, the heroine defies her assailant in terms very close to those of the Christian martyrs:

    commaunde what torments you please to bee provided, whether it be to be torne in peeces upon a Wheele, to bee whipped with scourges, to be burnt with fire, it will seeme to you to bee a newe kinde of sight: for one woman alone, will strive against so many punishments & depart the conqueror. [. . .] Wherefore, provide you whippes, a wheele, fire, iron [. . .] I am both naked, alone, and a woman: and have no defence, except my liberty, which can neither be whipped with rods, nor cut with iron, nor burnt with fire: that will I never leese, and if you cast me into the middle of the flame: there will not bee force inough therein to take it from me. (19–20)

20. Translated into English by Richard Hyrde, it was reprinted eight times by 1592 (Klein 67–69).

21. Foxe comments tartly on the Roman Catholic practice of canonizing nuns:

only because of the vow of their chastity solemnly professed. Concerning which chastity, whether they kept it or no, little I have to say against them, and less to swear for them. But whether they so kept it or not, if this gift of chastity which they professed were given them of God, worthy small praise was it in them to keep it [. . .]. But this I will say, that although they kept it never so perfectly, yet it is not that which maketh saints before God, but only the blood of Christ Jesus, and a true faith in him. (1: 384)

22. Among the stories Foxe includes are those of the virgin martyrs Potamiena (1: 156), Cecilia (1: 168), Eulalia (1: 270–2), and Tarbula, "a maid of right comely beauty," who "did choose rather to die, than to betray either the religion of her mind or the virginity of her body" (1: 283). Foxe's marginal glosses draw the reader's attention to the "chaste and continent behaviour of Eulalia" (1: 270) and the "Example of maidenly chastity in Tarbula" (1: 283). Foxe also quotes at length, and with approval, Prudentius's erotically charged accounts of Agnes's martyrdom (1: 272–3). He even records the deaths of Ursula and her eleven thousand virgins: "many perished in the sea, [and] some were taken of the infidels marching upon the borders; by whom because they would not be polluted, all were destroyed . . ." (1: 313).

23. See Gerard's *Autobiography* (46–49) for an eyewitness account of the cult in the 1590s. An appendix to the volume (265–66) gives further details.

24. Jensen suggests that in the early Stuart period "vestigial Catholic culture was not so much tolerated as invisible, still so much a part of the early modern English world that only in unusual contexts did it seem objectionable" (10).

25. For the campaign to suppress the old saints' lives, see H. White 67–95. For the preservation of Catholic books see C. E. Wright 170, R. MacDonald 102 and Gasper, "Sources" 30–31. A considerable body of Catholic literature in English was published after 1558, secretly or abroad, including John Wilson's *English Martyrologie* (1608) and Villegas's *Flos Sanctorum.* The latter went through four editions between 1609 and 1628. These would have circulated among the English Catholic community. Shakespeare's patron, the Earl of Southampton, though not Catholic himself, came from a devoutly Catholic family: In 1605 Chamberlain reported "above two hundreth pounds worth of popish bookes taken about Southampton House and burned in Poules Churchyard" (1: 202). For a list of such works, see Allison and Rogers's catalogue.

26. Unless otherwise noted, the dates given for plays are those found in the *Annals of English Drama.*

Chapter 2: Latter-Day Saints

1. I accept Edwards's judgement, put forward in his New Penguin edition, that *Pericles* "was conceived as a whole by a single mind," but was probably exe-

cuted by Shakespeare and a collaborator (39). However, I refer to the author(s) simply as "Shakespeare."

2. Hoeniger treats the relation of *Pericles* to the miracle play in the Arden (lxxxviii-xci). See also Love 127–33 and Felperin.

3. Negotiations for the Infanta began as early as 1614 (Akrigg 324), but the period between 1618 and 1624 marks the acme of Spanish influence in the Jacobean court (Mattingly 245–55). The repeal of anti-Catholic legislation was persistently brought forward by Spain as a condition of marriage (Akrigg 324–27, 336–60). In response to this pressure there was a general relaxation of penalties for recusants—see, for example, Gardiner 3: 135–37, 282, 345—though no change in legislation.

4. *The Second Maiden's Tragedy* (1611) also belongs to this group. As Lancashire demonstrates in her article of 1974, the play draws heavily on the hagiographic tradition. However, because of its strong affinities with the "Lucrece" plays, I discuss it in chapter 4.

5. The three later plays—all probably staged at the Red Bull—use the break between acts to heighten interest: A sexual threat is announced, but not effected, at the end of act 3 or 4, so that the audience has to wait through the interval in suspense. For the staging of *The Martyred Soldier* and *The Virgin Martyr* see Bentley 5: 1061 and 3: 265. The flyleaf (verso) of *The Two Noble Ladies* asserts that it was "oftne tymes acted with approbation At the Red Bull" (3). According to Gurr, the practice of breaking off the performance between each act was common to both public and private playhouses after 1607 (*Shakespearean Stage* 160–61).

6. For a discussion of sources see Bullough 6: 349–74 and Hoeniger's introduction to the Arden, xiii-ix. Citations of Gower and Twine are from Bullough 6: 375–482.

7. 1 Chorus.23–30. All citations of the text are from the New Penguin.

8. *"Scelere vehor, maternam carnem vescor, quaero fratrum meum, meae matris virum, uxoris meae filium: non invenio"* (Kortekaas RA 4.10–12).

9. Neely notes the change in voice, but arrives at different conclusions (168–69).

10. The motif was probably adapted by the hagiographers from Greek romance. It appears in Plautus's *Rudens* (c. 200 BCE) in which the heroine is, like Tharsia/Marina, kidnapped by pirates and sold to a pimp before being providentially delivered and reunited with her father. It also appears in Xenophon's *Ephesian Tale,* a romance of the second or third century, perhaps related to the Apollonius legend (Kortekaas 130). In Xenophon's romance, however, chastity is a sentimental rather than a moral issue, and when the heroine finds herself, like Tharsia/Marina, in a brothel, she feigns epilepsy and so chills the desire of the patrons.

11. Since, as I argue above, the legends of the saints continued to circulate in post-Reformation England, hagiographical knowledge does not imply Catholicism. However, Brownlow's discussion of "John Shakespeare's Recusancy" suggests that William Shakespeare may indeed have been raised in a Catholic home.

12. Judges are commonly driven into a frenzy by the virgins' eloquent challenge to their authority. Eufemia's judge was "much confounded, so that he almost died for anguish and sorrow" (De Voragine 5: 146). See also De Voragine's accounts of Lucy (2: 135–36), Agnes (2: 248), Juliana (3: 49), Barbara (6: 202–3) and Cecilia (6: 252). Christopher St. John draws attention to the parallel with the suffragettes in her introduction to Hrotsvitha's plays (xix).

13. Mullaney notes the shift from Tharsia's sentimental performance to Marina's sermons but arrives at different conclusions (*Place* 143–45).

14. See Edwards's remarks in the New Penguin *Pericles*, 21–26.

15. Marina's riddling here clearly stems from Tharsia's riddling skill in the *Historia*. In the *Historia*, however, as in Gower, the *Gesta* and Twine, Tharsia displays this power only when she rescues her father. There is no hint of it in the brothel.

16. *Vergine chiara e stabile in eterno,*
    Di questo tempestoso mare stella,
    D'ogni fedel nocchier fidata guida,
    Pon' mente in che terribile procella
    I'mi ritrovo sol, senza governo. [. . .] (Canzoniere CCCLXVI: 66–70)

    I use Warner's translation of this passage (*Alone* 263).

17. Given the connection I posit between Shakespeare's heroine and the Virgin Mary, the choice of *Pericles* for a Candlemas entertainment in 1609 in a recusant Yorkshire household (Sisson 138) seems particularly appropriate. Candlemas (February 2) commemorates the purification of the Virgin Mary after childbirth.

18. See Willis on Marina's "quasi-magical" power and her role in Pericles's healing (162–65).

19. Brownlow notes that Pericles's words echo:

    a well-known response that follows the second lesson in the Matins of the Virgin (a very popular devotion, known to every Tudor Catholic layman from his primer), translated in the vernacular primers as "Blessed art thou Virgin Mary, thou barest our Lord. Thou hast borne him that made thee." Shakespeare might have heard those striking words repeated in family devotions when he was a child. (190)

Shakespeare might have also encountered this paradox in the Protestant *Acts and Monuments,* where Foxe recounts at some length the story of Eugenia, a virgin of Alexandria who flees her home and assumes a masculine identity to pursue Christian learning. Like the biblical Joseph, Eugenia is slandered by a would-be seductress. Providentially, however, the judge before whom she appears is her father. She reveals herself to him and he, with her mother and brothers, is converted to the Christian faith. Thus, Foxe declares, "the father of her by nature, now *by grace was begotten of his own daughter to a more perfect life;* and whom once he thought to have been lost, not only he found again, but also with her found his own soul, and his own life, which before he had lost indeed" (1: 213, emphasis added). This hagiographic romance includes a number of elements present in Shakespeare's play: A beautiful and learned virgin, whose chastity is demonstrated in an episode of sexual shame; a providential reunion of father and daughter, with the father in a position of temporal power; and the transforming revelation of the daughter's identity, which brings the father from spiritual death to life.

20. Published in 1638, 11 years after Shirley's death, the Quarto is dedicated to Sir Kenelm Digby, a prominent Catholic courtier (*Dictionary of National Biography* 5: 965). Bullen reprints *The Martyred Soldier* in his *Old English Plays*. All citations are from this edition. The Quarto text is divided into acts only; Bullen adds scene divisions but no line numbers. References are thus to act, scene and page.

21. Buc's extraordinary tolerance may have been due to the general lightening of restrictions on Catholics at this time or to growing incompetence: In 1620 he was reported mad and in 1622 relieved of his duties for medical reasons (*Dictionary of National Biography* 3: 171).

22. Shirley's "Genzerick" appears as "Gaiseric" in modern histories. He ruled his tribe for almost 50 years (428–77 CE); his son Hunneric (Shirley's "Henericke") succeeded him in 477. See *The Columbia Encyclopedia* (1963) under "Gaiseric" and "Vandals." To avoid confusion I retain Shirley's spelling of the Vandal names.

23. Shirley belonged to an extraordinary Roman Catholic family: See the articles in the *Dictionary of National Biography* on his father, Thomas—courtier, pirate and MP—and his uncles, Anthony and Robert.

24. Arius (c. 256–336 CE) denied the equality of God the Son with God the Father. His teaching was condemned as heretical at Nicaea in 325 (Kelly 226–37). The Vandals, however, with the rest of the Germanic tribes, were converted to Christianity by Arians.

25. See Semper's account of a controversy between Thomas Leke, a Catholic priest, and his superiors. In March 1618 William Harrison, Archpriest of England, forbade priests to attend plays "acted by common players upon common stages" (45). Leke protested, claiming that "most of the principal Catholicks about London doe goe to playes . . ." (46). The prohibition was subsequently revoked (51).

26. Victor's *De Persecutione Vandalica* appeared as *The Memorable and Tragical History, of the Persecution in Africke: Under Genesericke and Hunericke, Arrian Kinges of the Vandals* (STC 24714). Buckland prudently does not acknowledge the translation by name. See, however, Allison and Rogers number 847 and the *Dictionary of National Biography* on Buckland.

27. Thus Buckland quotes *Ecclesiastes* 1: 9 as an epigraph: *"Nothing new under the Sunne."* Although in his preface he disclaims any intention of comparing the reign of Elizabeth with those of Genzericke and Henricke (8), in his final pages Buckland makes the parallel between Arianism and Protestantism explicit (211–12).

28. 39–42 and 97–98. In a third story, concerning "the manly courage of a Physicions wife" (116), a couple are imprisoned separately. The wife is told her husband has recanted, and is advised to do the same. She requests to see him. When they are reunited before their judges, she attacks and abuses him for his inconstancy, only to find that he has remained faithful (117–18). The point of all these stories is the Catholic's heroic indifference to earthly ties for the sake of the faith.

29. Shirley also wants to recall Lear and Cordelia. He has Bellizarius say to her, inconsequently:

> Wee'le live together, if it please the King,
> And tell sad Stories of thy wretched Mother;
> Give equall sighes to one anothers griefe,
> And by discourse of happinesse to come
> Trample upon our present miseries. (3.2; 212)

There is no reason why Victoria should be excluded from their life together, the subject only of "sad Stories"—she is not dead yet—except that her absence isolates the father-daughter bond and lends a specious strength to the Lear-Cordelia echo.

30. Almost all the virgin martyrs are imprisoned. Eufemia is starved for seven days but fed by an angel (De Voragine 5: 146); Dorothea starves for nine days, but she too is fed by angels (De Voragine 7: 43). According to Aelfric, Eugenia starves for twenty days but is fed by Christ (1: 49). Barbara is miraculously healed in prison (De Voragine 6: 203). Juliana is imprisoned after severe torture, but when she appears before the provost the next day, she is "so well guerished, and her visage so fair and so shining," that he marvels (De Voragine 3: 48).

31. This is Henericke's first expression of lust for Victoria and its appearance here suggests the influence of the Katherine story: Shirley's tyrant has to catch up, as it were, with his hagiographic prototype, who desired Katherine *before* committing her to prison. The image of Victoria dismembered also seems to be a transference—out of sequence—from Katherine's story. Katherine was torn apart on a wheel (and was often pictured with one): Her notorious dismemberment perhaps accounts for Shirley's (or Henericke's) curious choice of words. There is nothing in the plot to explain it.

32. Victoria's transfiguration here probably owes something to the apparition of the Lady in *The Second Maiden's Tragedy,* who emerges from her tomb *"all in white, stuck with jewels, and a great crucifix on her breast"* (4.4.42.s.d.). The hint of necrophilia here also suggests the influence of the earlier saints play: see below, p. 97–8.

33. All citations of the text are from *The Dramatic Works of Thomas Dekker* 3: 365–480, edited by Bowers. Critical consensus attributes all of act 1, 3.1, 3.2, 4.3, and 5.2 to Massinger; 2.1, 3.3 and 4.2 to Dekker. The remaining scenes (3.2, 2.3, 4.1 and 5.1) are, according to Hoy, "essentially Dekker's," though each contains "traces of Massinger" (*Introductions* 3: 193).

34. *The Virgin Martyr* may be a Protestant response to Shirley's intervention in
the debate about England's political-religious policy. See Adler 35–36, and
Gasper, *Dragon* 136–65. The play's religious content has been treated re-
spectfully and even enthusiastically by some critics: See, for example, Hunt
158, Price 94–96, Champion 108–14, and Hoy *Introductions* 3: 194–96.
However, Wickham observes astutely that *The Virgin Martyr* "bears a closer
resemblance to films like *Ben Hur* or *Quo Vadis* than to genuinely religious
plays" (2.1: 62n). Similarly, Peter Mullany concludes that the authors use religious conflict primarily for its emotional, sensational and theatrical effects
(55–71).

35. See Baring-Gould 2: 176–77; Aldhelm, *De Virginitate* [prosa] chapter 47
(*Opera* 300–2); and De Voragine 7: 42–47. For an extended discussion of
the play's sources see Gasper, "Sources" and Peterson. Hoy presents a clear
summary (*Introductions* 3: 179–88).

36. See, for example, the accounts in De Voragine of Margaret (4: 66–72),
Juliana (3: 45–50) and Barbara (6: 198–205). Juliana is beaten naked by
her father; Barbara's father abuses her, drags her by the hair, imprisons
her, instigates her torture and finally kills her himself. Warner situates this
motif of paternal persecution in the wider context of folklore ("The
Wronged Daughter").

37. See Stallybrass's discussion of the female body as "patriarchal territory" in
early modern England.

38. Given the close association of female hair and sexuality in Western culture,
it is not surprising that, according to Wolfthal:

> In medieval art disheveled hair was a sign of rape and a man touch-
> ing a woman's hair implied the threat of sexual violation [. . .]
> [T]his motif recurs in numerous fifteenth-century woodcuts from
> the *Golden Legend* and the *Lives of the Saints* that show innocent
> women, such as Sts. Agnes, Thecla, and Euphemia, threatened with
> rape. Furthermore, later images of war, from the fifteenth through
> the seventeenth century, frequently employ the motif of a man
> pulling a woman's hair as a sign of sexual violation [. . .]. (70)

See also note 36, above. Commenting on a line in Euripides' *Iphigenia in
Aulis,* Joplin remarks, "The suggestion of a rape in the woman dragged by
her hair and screaming is unmistakable" (40n30).

39. Ford echoes Chapman's stage imagery in an analogous scene of marital violence in *Tis Pity She's a Whore* (4.3), when the jealous Soranzo enters *"unbraced, and* ANNABELLA *[is] dragged in."* "Thus will I pull thy hair," he
declares, "and thus I'll drag / Thy lust-be-lepered body through the dust"
(60–61). Marsden notes the use of disordered hair as a sign of rape in
Restoration drama but links it specifically to the presence of female actors:
"As the joint appearance of actresses and scenes of rape indicates, rape becomes possible as theatrical spectacle only when visible signs of the female
are present: breasts, bare shoulders, and 'ravished' hair" (186). Clearly,

however, as these scenes from Marston, Chapman and Ford demonstrate, "visible signs of the female" can operate powerfully without female actors.

40. Theophilus's staged possession and repossession of his daughters seem to me strongly incestuous. Significantly Massinger echoes this scene of paternal anger in *The Unnatural Combat* (1624–25), when Malefort, in a similarly public display of authority and impelled by incestuous passion, threatens to drag his daughter home by the hair (3.4.81–83). For a contrasting assessment of Theophilus, however, see Otten 143.

41. The use of the enchanted cave as a metaphor for the virginal body strongly suggests the influence here of Rowley's *All's Lost By Lust* (1619–20): See below, pp. 107–10.

42. Cf. Laura Brown's discussion of violence and pleasure in Otway's *The Orphan* (1680): "The female victim's pain is the obvious objective of the plot. But pain is connected not only with female sexuality, but, perversely and masochistically, with 'pleasure' as well" (71).

43. To say this is not to deny the metaphorical significance that, according to Gasper, we should attribute to Dorothea's virginity. Dorothea's "refusal of Antoninus's legitimate advances [. . .]," she argues, "symbolizes the refusal of the True Church to enter into non-confessional alliances" (158). While Gasper acknowledges that the "rape metaphor can seem in some respects rather distasteful," she declares that the "playwrights would not have used it without serious metaphorical purpose in mind: [. . .] virginity is the symbol of autonomy" (158–59). Indeed, this symbolism is integral to the hagiographic tradition. However, "serious metaphorical purpose" does not preclude an erotic and sensational treatment of sexual violence; nor does the authors' putative religious sincerity.

44. Its text survived in a manuscript that was clearly the author's fair copy and prompt-book (MS. Egerton 1994, fols. 224–244). Rhoads discusses the play's date in her introduction to the Malone Society reprint (*Two Noble Ladies* vi-vii).

45. This is the description offered, apparently with a view to publication, in another hand on the verso of the flyleaf (vii).

46. See De Voragine 5: 165–47; Aldhelm, *De Virginitate* [carmen] 1842–82 (*Opera* 429–30); Baring-Gould 10: 386–89; and *Bibliotheca Sanctorum* 3: 1281–85. Jacobean readers might have known Foxe's brief account: "Cyprian likewise, a citizen of Antioch, who, after he had continued a long time a filthy magician or sorcerer, at length was converted [. . .]. This Cyprian, with Justina a virgin, suffered among the martyrs" (I: 267–68).

47. An anonymous play, performed by the Children of the Chapel, some time between 1576 and 1594. See the account given by Brawner in the introduction to his edition of the 1594 text (*Wars of Cyrus* 10–20). As a comparison of the play with its historical source—Xenophon's *Education of Cyrus*—shows, this scene of magical assault was added to the story of Panthea and Araspas by the author. See *Wars of Cyrus* 20–28 and 134–35n.

0

48. The close connection between lust and murder in Desdemona's killing is suggested by the influence of the scene on Goffe's *Courageous Turk* (1618). In Goffe's source the king murders his concubine in a public hall (38); in the play he murders her as she lies sleeping in bed. Before the murder he meditates on her beauty, deliberating between killing her and consummating their marriage (2.3–5).

49. A connection between Cyprian and Cerimon here is suggested by a verbal echo. Gazing at Thaisa as she is about to awake, Cerimon says:

> Her eyelids, cases to those heavenly jewels
> Which Pericles hath lost, begin to part
> Their fringes of bright gold. The diamonds
> Of a most praisèd water doth appear
> To make the world twice rich. (3.2.97–101)

At the parallel moment, Cyprian says:

> If she did unlidde
> those yuory cases, 2. rich diamonds
> would dazle humane eies, and tell the world
> earth is too poore to buy them. (1783–86)

The benevolent physician/magician Cerimon is of course diametrically opposed to Cyprian in this scene. However, in the Latin source of *Pericles* this resurrection scene is erotically charged: There is indeed a strong suggestion of symbolic sexual violation. In the *Historia*, the *Gesta* and Twine, the princess is recovered by Cerimon's assistant, who removes her clothes to anoint her with oil and feels warmth in her breast. He then carries her to his own bed, where, to instil heat, he lies on top of her, chafes her, and rubs her with warm oil. When she awakes her first words are a defence of her chastity (Kortekaas RB 26–27). Something of this sexual danger lingers in the much more decorous Shakespearean scene. On waking to find herself surrounded by three strange men, Thaisa cries, "O dear Diana!" (3.2.104).

50. For a fuller discussion of these scenes in *Cymbeline* and *The Second Maiden's Tragedy*, see below, pp. 87–9 and 97–8. Field's *A Woman is a Weathercock* offers another instance of this topos: Scudamore enters Bellafront's chamber and finds her *"asleep in a chair, under a taffata canopy"* (3.2: p. 378). Before waking her, he reflects on her beauty and perfidy: In the subsequent dialogue he threatens her with death (p. 379).

51. Justina's perilous situation—immobile in a chair, surrounded by a lustful magician and attendant demons—in many respects anticipates that of the Lady in Milton's *Mask* (immobile in a chair and surrounded by a lustful magician and his attendant *"rabble"* [658.s.d.]): The Lady, however, is conscious and verbally active.

52. This raises the vexed question of gendered response to representations of sexual violation and pornography, and the problem of the viewer (of either sex) who enjoys identifying with the victim. There has been much critical debate

about "the male gaze" and that of the female spectator in the last 20 years; Modleski provides a useful survey of the discussion (1–15).

53. This topos, of course, becomes a staple of Gothic fiction and, later, film: Stoker's *Dracula* is the *locus classicus*. Pictorially, Fuseli's "Nightmare," in which a demon crouches on the distended body of a sleeping woman, provides a powerful variation on the theme. As Nicolas Powell observes, "there can be little doubt that the girl in Fuseli's painting is experiencing an imaginary sexual assault" (60). See Powell for a discussion of the painting's sources, analogues and influence.

54. On eye magic see Woodbridge (*Scythe* 60–67): As she argues, Tarquin's aggressive gaze in Shakespeare's poem is related to evil eye beliefs; thus Tarquin is endowed "with the gaze of the best known of evil-eye possessors, the basilisk, or cockatrice: [A]s Lucrece pleads, he fixes her 'with cockatrice' dead-killing eye' (540)" (*Scythe* 62). Similarly, in *Women Beware Women* Bianca describes her rape by the Duke in terms that suggest the evil eye:

> Now bless me from a blasting; I saw that now
> Fearful for any woman's eye to look on.
> Infectious mists and mildews hang at's eyes,
> The weather of a doomsday dwells upon him.
> Yet since mine honour's leprous, why should I
> Preserve that fair that caused the leprosy?
> Come poison all at once. (2.2.420–26)

Bianca thus figures her sexual violation as a contaminating penetration of her vision that leaves her "leprous." According to early modern optical theories, "like sex, the act of looking involved a penetrating emission of vital spirit and also opened the looker to potentially contaminating penetration by what was seen" (Crane 287).

55. See above, p. 29.

Chapter 3: The Classical Paradigm: Lucrece and Virginia

1. In addition to a variety of classical and medieval accounts and the versions discussed below, both stories were included in Painter's *Palace of Pleasure* (1566). Pettie adapted the story of Virginia for his *Petite Pallace* (1576), while Lucretia was the subject of ballads (SR 1568 and 1569), Middleton's *The Ghost of Lucrece* (1598) and a lost Latin play at St. John's College, Oxford (1605). Sidney Lee surveys the various versions of the Lucretia story in his introduction to the facsimile edition of Shakespeare's poem (10–11).

2. See Waith's introduction to the Oxford edition for the problems of dating *Titus* (4–11). He concludes that "it was first performed in the years 1590–2, and was revised for the first recorded performance of January 1594" (20).

3. In his edition of Webster's *Complete Works* Lucas dated the play 1625–27 (3:121–30), and Bentley cautiously approved (5: 1247). Since then, how-

ever, Steppat has found evidence that points to a date before 1616. Since the last lines of the play make such pointed reference to the Lucrece story (5.2.186–95), I suggest that Webster, prompted by the success of Heywood's *Rape of Lucrece*, wrote the play sometime after 1608 and before 1616— probably before *The White Devil* (1613) and *The Duchess of Malfi* (1614).

4. Bk. 1, ch. 1. In Holland's 1600 translation Livy's condemnation of contemporary vice sounds authentically Jacobean: "Now of late daies wealth hath brought in avarice, abundant pleasures have kindled a desire by riot, lust and loose life, to perish and bring all to naught" (2–3). For Livy's context, see Joshel 11–15.

5. For a feminist analysis of Livy see Joshel. For Livy's art see Ogilvie's commentary 218–29; 476–94 and 503–6. Ogilivie praises in particular "the skill and poignancy with which L[ivy] has constructed what is one of the noblest episodes in his narrative. Verginia was for him a supreme example of the virtue of *pudicitia*, a supreme condemnation of *libido*. [. . .] Yet for all its beauty, the story of Verginia is entirely devoid of historical foundation" (476–77). The enthusiasm of a modern scholar—Ogilvie published his commentary in 1965—indicates the remarkable persistence of the legend's appeal. See Donaldson and Jed for the transmission of the Lucrece myth in western culture.

6. For Rome as a legendary site of exemplary virtue in Renaissance England, see Kahn 1–26; for Lucretia and Virginia as models of female excellence in this period, see Pearse 71 and 90–99. Donaldson discusses both the sanctification of Lucretia and traditions of dissent (25–39). Drawing on marginalia to demonstrate dissident reader responses to Lucretia in early modern England, Sasha Roberts analyzes the ways in which later quartos of Shakespeare's poem (Q6-Q9) "reiterate the conventional, mainstream interpretation of Lucrece as an icon of marital chastity and coerce a moralistic reading of the poem" (131).

7. Significantly in *The Martyred Soldier, The Virgin Martyr* and *The Two Noble Ladies* the dramatists compensate for this limitation in their hagiographic material by adding a war story.

8. Livy spells the names of the principal characters "Appius," "Verginius," and "Verginia." Like Chaucer, R. B. spells them "Apius," "Virginia" and "Virginius." Like Webster, I retain Livy's spelling of the former ("Appius"), but use the anglicized "Virginia" and "Virginius."

9. Citations are from Skeat's edition of Chaucer's *Works* 4: 290–98. Chaucer's source was *Le Roman de la Rose* (5613–82): See Skeat's appendix for an account (5: 435–37).

10. Shakespeare deviates from Livy's version of the story by having the father kill his daughter *because* she has been raped, rather than to prevent her rape. Norgaard points out (139–40) that Shakespeare may have known the alternate tradition reported by Lodowick Lloyd in *The Pilgrimage of Princes* (1573). In any case, Lavinia's rape enhances her likeness to Lucretia. Hunter

discusses Shakespeare's general debt in *Titus* to the early books of Livy (182–88).

11. All citations of the text are from Waith's Oxford edition.

12. Anna Calder-Marshall, who acted Lavinia in the B.B.C. production, compared the play to a "video nasty" (B.B.C. *Titus* 22)—a phrase that aptly suggests the brutality of the violence.

13. Kahn observes that "Titus's subjectivity as a tragic hero is generated from his daughter's pain inscribed as his own" (48). See also Douglas Green 322.

14. In a discussion of Philomela in the Renaissance, however, Lamb observes that "Shakspeare's play consistently softens Lavinia's role to portray her more as a figure of pathos than of anger" (219).

15. On Lavinia's silence, see Kendall (314–15), for whom the girl embodies "the mysteries of language" (314), Fawcett 265–66, Douglas Green 323–26, and Gajowski.

16. Citations of the text are from the Arden edition of Shakespeare's *Poems* by F. T. Prince.

17. Shakespeare's main sources are Ovid's *Fasti* (2: 721–852), Livy in Painter's translation (the Second Novel in *The Palace of Pleasure*) and Chaucer's *The Legend of Good Women* (1680–1885). They are reprinted in the Arden edition of Shakespeare's *Poems* (189–201) and Bullough 1: 184–202 (Bullough also includes Gower's translation of Ovid). See Bullough 1: 179–83 for a brief discussion of the sources.

18. Empson suggests that the reader of the poem may be guilty, "having taken a sexual pleasure in these descriptions of sexual wrong" ("Narrative Poems" 11).

19. As Nancy Vickers points out, Tarquin is first provoked not by Lucrece's beauty but by Collatine's praise of it and of her chastity (212–13), and there is thus a large element of male rivalry in the crime. Once he sees Lucrece, however, Tarquin is "enchanted" by her beauty (83), and, in spite of his conscience, unable to stop himself (288–94).

20. Empson believes that Shakespeare's Lucrece "took an involuntary pleasure in the rape [. . .] that is why she felt guilty, and why some of her blood turned black [. . .]" ("Narrative Poems" 11). He might also have cited Lucrece's fear of pregnancy (1062–64), since a woman's sexual pleasure was widely believed to be necessary for conception (McLaren 20–21). Empson's hypothesis is interesting because it highlights the persistent male myth that rape is a source of erotic pleasure for the victim. Cf. Augustine on rape: "[A]lthough it does not destroy a purity which has been maintained by the utmost resolution, still it does engender a sense of shame, because it may be believed that an act, which perhaps *could not have taken place without some physical pleasure,* was accompanied also by a consent of the mind" (*City of God* bk. 1, ch. 16; Penguin 26; emphasis added). Donaldson notes that speculation about Lucrece's guilty pleasure was "common" among Christian writers and cites Coluccio Salutati and Bandello in particular (36). Baines argues that, given the early modern legal dictum that conception means consent and

thus negates the accusation of rape, Lucrece's fear of pregnancy would prompt Shakespeare's "gentlemen readers" to believe (like Empson) that Lucrece did experience an involuntary pleasure and that this possibility "functions as a prurient secret at the heart of this text—a male fantasy of totalizing phallic power, highly reassuring to patriarchy" (88–89).

21. Prince (Shakespeare, *Poems* 111) aptly connects these curses with those of the aggrieved women in *Richard III* (1.2.112, 1.3.223f., and 5.3.118f.), and the comparison illuminates some of Lucrece's rather sinister power. Middleton picks up and expands the vengeful aspect of Shakespeare's Lucrece. The ghost of *his* Lucrece returns from hell to rehearse her woes, impelled by "Medea's magic, and Calipso's drugs, / Circe's enchantments, [and] Hecate's triform" (*The Ghost of Lucrece* 1–2).

22. As Kahn observes (28), critics have been disturbed by Lucrece's copious rhetoric. She cites, in particular, Prince's comment: "After her violation, Lucrece loses our sympathy exactly in proportion as she gives tongue" (Shakespeare, *Poems* xxxvi). While I wish to distinguish my position from Prince's, his loss of sympathy for Lucrece seems to me to emerge from the poem's subtle demonization of her after the rape, when her grief is "mad and too much talk affords" (1106). Berry, however, argues that "Lucrece is represented [. . .] as an important but unorthodox example of Renaissance virtù" (34).

23. Performed by Queen Anne's company at the Red Bull (1607). Citations of the text are from *The Dramatic Works of Thomas Heywood* 5. Since the play is not divided into acts, scenes or lines, references are to pages.

24. Indeed, the values embodied by Lucrece are wholly consistent with those celebrated in Heywood's *A Woman Killed with Kindess*, where the exemplary Susan nearly kills herself to preserve her chastity, and the adulterous Anne dies penitent.

25. Tarquin's deliberations before the crime (217–22) read like a digest of those in *Lucrece* (190–280); as in the poem, so here Tarquin draws the curtains and is dazzled by the beauty of the sleeping woman (222); Lucrece attempts to dissuade him using the same arguments as her predecessor; later her dialogue with her maid (234–35) recalls Shakespeare's use of the maid (1217–39) and the groom (1338–58).

26. The following might be construed as having political significance: *"When Tarquin first in Court began"* (179); *"Let humor change and spare not"* (180); *"Lament Ladies lament"* (181); *"Why since we souldiers cannot prove"* (181–82); *"Though the weather jangles"* (199–200); *"Come list and harke"* (230); *"I'de thinke my self as proud in Shackles"* (231–32).

27. *"Now what is love I will thee tell"* (180); *"The Gentry to the Kings head"* (198); *"O yes, roome for the Cryer"* (213); *"The* Spaniard *loves his ancient slop"* (216).

28. Shakespeare places his lyric, "Hark, hark, the lark at heaven's gate sings," in a parallel position in *Cymbeline* (c. 1609): Cloten has it sung outside Imogen's window the morning after Iachimo has symbolically raped her (2.3.19–26). Since, according to Holaday's introduction to *The Rape of Lucrece* (31),

"*Packe cloudes away*" does not appear until the fourth edition of 1630, Heywood seems to have been influenced by Shakespeare here.

29. Underneath the title the publisher advertises "the severall Songs in their apt places, by *Valerius* the merry Lord among the Roman Peeres. The Copy revised, and sundry Songs before omitted, now inserted in their right places" (reprinted in *Dramatic Works* 5: 161).

30. Helms, however, reads this song as "Heywood's strategy for subversion" of "the chivalric ideology that makes the aristocratic female body a boundary marker and its unlawful penetration a pretext for civil wars" (562). She nevertheless also observes that this strategy "authorizes a randy trio of goons to mock the victim of sexual violence" (562).

31. "*Alarum, fight with single swords, and being deadly wounded and painting* [sic] *for breth, making a stroak at each together with their gantlets they fall*" (252).

32. Introduction to *Appius and Virginia* in *The Complete Works of John Webster* 3: 146. All citations of the text are from this edition. The play has been attributed by some to Heywood: Lucas concludes that there is a strong probability that Heywood did have a hand in it, but attributes it primarily to Webster (145). We know nothing of its early stage history (Bentley 5: 1246–48).

33. In addition to Livy's account, which he probably read in Holland's translation, Webster used *The Roman Antiquities* of Dionysius Halicarnassus, book 11 (vol. 7 of the Loeb *Dionysius*, 2–155), available both in the original Greek and a Latin translation. Painter's version of Livy appears in the Fifth Novell of the First Book. For a brief discussion of the sources see Lucas's introduction to the play (Webster, *Complete Works* 131–33).

34. According to Roman law a father had the right to kill his child: W. V. Harris observes that "no ancient writer suggests that his [Virginius's] action was illegal," and cites a parallel case in which a father killed his daughter *after* she was raped (87).

35. Abraham and Jephthah provided biblical models of pious parricide, however.

Chapter 4: "Some Injury in a Matter of Women": Variations on the Classical Theme

1. *The Revenger's Tragedy* (SR 1607) is probably earlier than either *The Rape of Lucrece* or *Appius and Virginia*. The title page of the 1607/8 Quarto assigns the play to the King's Men. The authorship is still disputed, but the weight of the evidence points to Middleton rather than Tourneur (Murray 144–73; Lake 136–49; MacD. P. Jackson's introduction to the Quarto facsimile, attributed to Middleton, 15–31). All citations of the text are from the New Mermaids edition by Brian Gibbons.

2. Like the queens in *Titus* and *Cymbeline,* the Duchess has married the ruler as a widow; like them she has a son by a prior marriage whose sexual aggression has political consequences. According to Livy, Tullia also married Tarquin as a widow (bk. 1, ch. 46–47).

3. "It is better to die virtuously than to live dishonored."
4. The lords swear to execute the rapist in court if a corrupt justice delays his punishment again (61–64), but in the event he is sentenced to death and accidentally executed (3.4.26–78).
5. For a contrasting view of Vindice's trick see Robertson, who finds in it a "positive, though threatening, representation of feminine agency" (229). She argues that "although the female figure has no tongue, cannot speak, and is manipulated by the male character—thus in one way the perfect image of the woman as male property—possession is explicitly returned to the female murder victim. The scene offers a terrifying image of the rigorous execution of justice by the victim herself" (228). Steven Mullaney, however, observes that Vindice's trick subjects "his chaste lover [. . .] to the fate she died to avoid: she is the painted lady, the courtesan, the whore" ("Mourning" 160).
6. See Nosworthy's introduction to the Arden *Cymbeline* (xiv-xvii). All citations of the text are from this edition.
7. *Decameron,* the ninth novel of the second day. The English prose tale *Frederyke of Jennen,* translated from the Dutch, was also a source for the wager plot. Both are reprinted in Bullough 8: 50–78. For Shakespeare's use of his sources see also the Arden xvii-xxviii, Sexton 61–76 and Swander.
8. For Imogen's national significance, see Woodbridge, *Scythe* 55–56, and Mikalachki 108–14.
9. For this reason Imogen's virginity has been the subject of critical contention. Gesner argues that the marriage must be a "handfasting, an old form of the irregular or probationary marriage" (102) and as yet unconsummated, but Kirsch disagrees (148). There is no way to resolve the dispute: Shakespeare withholds all information on the lovers' history.
10. Amoret is abducted from her wedding feast by the "vile Enchauntour Busyran" (*Faerie Queene* 4.1.3) to endure captivity and multiple threats to her chastity. Imogen shares her liminal status with other sexually threatened heroines on the Jacobean stage: For example, Marston's Sophonisba, the Lady in *The Second Maiden's Tragedy* (see below, p. 94) and Castabella in *The Atheist's Tragedy.* Dubrow notes a similar ambiguity about Shakespeare's Lucrece: The poem's imagery suggests that she is "virtually virginal despite her marriage" ("Mirror" 412). The peculiar excitement and anxiety generated by Imogen's anomalous position extend beyond the play to its critical reception and may partly account for the cult of Imogen that flourished in the nineteenth century (see below, n. 14).
11. The role of Cecropia and Amphialus in *The Countess of Pembroke's Arcadia* presents another parallel. Sidney even transfers most of the guilt for the sexual threat to Philoclea from the son to the mother (see, for example, Cecropia's speech at bk. 3, ch. 17 [532–34]).
12. Imogen's prayer ("To your protection I commend me, gods" [2.2.8]) also echoes that of Heywood's Lucrece as she goes to bed ("*Iove* unto thy protection I commit / My chastitie and honour to thy keepe" [220]).

13. Miola lists many close verbal parallels between *Lucrece* and *Cymbeline,* including the imagery of siege and invasion, the walled fortress and sacred temple, locks, treasures, flowers and even the "azure" colour of the heroines' veins (52–53). Curiously, Miola also compares Iachimo and Tarquin to the latter's advantage:

> Iachimo fancies himself another Tarquin and Shakespeare delights
> in fostering the illusion. All the while, however, the disparity be-
> tween the brutal rape and the sneakily malicious note-taking comes
> into effect. Tarquin violates Lucrece, her household, family and city;
> Iachimo merely plays a cheap trick. (53)

> Miola's evaluation of the respective crimes suggests that he accepts what
> Brownmiller calls, "the myth of the heroic rapist" (283–308). Similarly,
> Bergeron argues that Iachimo's "passive" response to the sleeping Imogen re-
> veals his impotence (165)—as if a *real* man would have been unable to re-
> frain from an "active" rape. Bergeron's judgement is echoed by Robert
> Lindsay, who played Iachimo in the B.B.C. production: "I believe he's im-
> potent: he's so warped inside he isn't able to confront a woman honestly, face
> to face" (B.B.C. *Cymbeline* 24).

14. In his 1886 edition, Ingleby reacts violently to the suggestion that Iachimo actually kisses Imogen: "Capell's vulgar interpretation is too monstrous to need refutation. Shakespeare could not have intended the profligate Italian to sully the purity of Imogen's lips. He does *not* kiss her" (59).

15. A possible tragic variation would be a wasting sickness, like the one that kills Penthea in Ford's *The Broken Heart,* or that which consumes, at much greater length, Richardson's Clarissa. A comic variation would be Imogen's marriage to Iachimo (see chapter 5, below) although her prior marriage to Posthumus, of course, precludes this "happy" ending.

16. Nosworthy notes the liturgical echo without comment (49).

17. In the B.B.C. production Elijah Moshinsky picked up and magnified the suggestion of Imogen's complicity:

> Iachimo so disturbs Imogen, who may or may not be pure, that she
> in a sense has a nightmare about the presence of Iachimo which we
> know, objectively, to be true. [. . .] On television you get the op-
> portunity of actually making the scene like her nightmare [. . .] we
> do all the filmic techniques of close-up and time-lapse and silhou-
> ettes and menacing shots and the suggestion of his nakedness, so he
> has a rather potent sexual force. (B.B.C. *Cymbeline* 17)

> Iachimo's evil, Moshinsky observes, "actually depends on having made the
> other person guilty. The removal of the bracelet is in fact a rape. It's an ex-
> traordinary scene, the rhythm of it is so sexual" (23). Robert Lindsay con-
> curs: "I think the trunk scene is very pornographic [. . .] He's excited, he's
> been in this trunk for hours, he's hot and sweaty and he gets out and he can
> do anything he wants; and Imogen is dreaming about being raped [. . .]"

(B.B.C. *Cymbeline* 24). Robin Phillips's 1986 production at Stratford, Ontario enhanced the pornographic aspect of the bedroom scene. According to Roger Warren,

> Innogen seemed in greater danger than she has seemed in any other modern production, especially when Giacomo drew the bedclothes completely off her and straddled her before kissing the mole on her breast [. . .]. As she twisted and turned, she appeared very disturbed [. . .]. This graphic staging implied much more strongly than Moshinsky's that the near-rape was an externalizing of her nightmare. It also carried another suggestion. Giacomo wore jodhpurs, which gave the additional impression of 'riding' her in the sexual sense; and as her body twisted and buckled beneath his, the distinction between near-rape and actual rape began to seem very thin. [. . .] Innogen seemed much more soiled than usual, far more than in the television version, for all Lindsay's nakedness. (*Cymbeline* 91)

18. Cf. the false Una that enrages Spenser's knight (*Faerie Queene* 1.1.36–2.6) and the pseudo-Amoret in Fletcher's *The Faithful Shepherdess* who provokes Perigot to wrath by her loose behavior (3.1.243–318; *Dramatic Works* 3). Like Posthumus, Perigot attempts to kill his betrothed and nearly succeeds twice (3.1.318, 344 and 4.4.162).

19. Siemon observes:

> In its particular details Posthumus's plan differs from Cloten's, although the degree of distinction one allows depends upon one's readiness to take as a phallic symbol the sword which is to be the instrument of Posthumus's revenge. At the least, both respond with violence to sexual humiliation and just as Shakespeare places the scenes of humiliation back to back, so he puts together the scenes in which the plans for violent revenge are most fully stated. The invitation to compare is patent, and the comparison shows the disintegration of Posthumus's character as a process by which he adopts Cloten's manners and some, at least, of his morals. (59)

20. Siemon, Kirsch (154–59), and Warren, "Theatrical Virtuosity," are particularly good on the relationship between Posthumus and Cloten.

21. If, as seems likely, Shakespeare wrote "Innogen" rather than "Imogen," the heroine's name expresses her essential purity. See Warren, *Cymbeline* viii.

22. For the unjust punishment of sexual crimes in hagiography, see, for example, the legend of St. Marine (De Voragine 3: 226–28). Jardine relates this motif to the Grissill legend and identifies it as one of "the saving stereotypes of female heroism" in the Renaissance (182–83).

23. At 3.4.54, 65–66, 68, 72, 75, 79, 97, 102.

24. Thompson notes the similarity between Imogen in this scene and Ovid's Philomela, who when she "sawe the sworde, she hoapt she should have dide, / And for the same hir naked throte she gladly did provide" (*Metamorphoses*

6. 705–6, Golding's translation; Thompson 24). The connection strengthens the sense of Posthumus's attack on Imogen as a form of sexual aggression, like Tereus's assault on Philomela.

25. Sexton cites as an analogue to *Cymbeline* a French miracle play, *Miracle d'Oton, Roi d'Espagne,* that makes similar use of a wager plot: There the Virgin reveals the truth and effects a happy ending (62–73). Sexton observes that Imogen and the heroine of the French play "suffer with strength and dignity, but their helplessness is emphasized. They are exhausted and heartsick" (73).

26. Since the wicked Queen dies of a "fever with the absence of her son" (4.3.2), we may credit Guiderius and Imogen with that death as well.

27. Cymbeline graphically insists on their national significance when he calls the unknown heroes, "the liver, heart, and brain of Britain, / By whom [. . .] she lives" (5.5.14–15).

28. The role of the heroine in Greene's *James IV* presents an interesting analogue. There the lustful King of Scotland attempts to have his wife murdered. Disguised as a boy, the Queen flees to the forest with a faithful retainer and is wounded. Providentially she survives to reconcile her husband with her father, the King of England, and prevent national disaster.

29. Schoenbaum lists many similarities between *The Revenger's Tragedy* and *The Second Maiden's Tragedy* (186–87), and there is strong evidence that both were written by Middleton. For date and authorship of *The Second Maiden's Tragedy* see Lake 185–91, and Lancashire's introduction to the Revels edition, 14–23. All citations of the text are from the Revels.

30. For the play's sources, see Lancashire's article, "*The Second Maiden's Tragedy:* A Jacobean Saint's Life;" for the debt to earlier versions of the Virginia story, 273–76. See also the introduction to the Revels, 23–32, and Appendix C, which reprints relevant passages from the sources (289–302). Sophronia's story is first reported by Eusebius in his *Ecclesiastical History:*

> when she learnt that at her house were those who ministered to the tyrant in such deeds (and she also was a Christian), and that her husband, and he too a prefect of the Romans, through fear had permitted them to take and lead her off, she begged to be excused for a brief space, as if forsooth to adorn her person, entered her chamber, and when alone transfixed herself with a sword. And straightway dying she left her corpse to her procurers; but by deeds that themselves were more eloquent than any words she made it known to all men, both those present and those to come hereafter, that a Christian's virtue is the only possession that cannot be conquered or destroyed. (bk. 8, ch. 14; Loeb 2: 311)

> Foxe summarizes Eusebius's account and concludes by observing that when the tyrant learned of the woman's death, he was "so far past shame, that instead of repentance, he was the more set on fire in attempting the like" (1: 247).

31. As Weller and Ferguson note in their introduction to Cary's play (5–6), *The Tragedy of Mariam* (1602–09; published 1613) probably exercised a minor

influence on *The Second Maiden's Tragedy*. For the story of Drusiana see above, p. 29.

32. For the relationship between the two plots see R. Levin's "Double Plot," and Lancashire's introduction to the Revels 33–34.

33. Lancashire calls attention to the significance of the stage imagery (*Second Maiden's* 89, 100).

34. Subsequently, though in a different spirit, the Lady's father pays a similar tribute to her "princely carriage and astonishing presence" (2.1.81).

35. It is perhaps no more than a coincidence that in Hrotsvitha's play *Callimachus*, Drusiana replies to the blandishments of her pursuer with the declaration, "I shall not change, be sure of that" (54). Elizabeth, the virgin queen, used the motto "Semper eadem," "always the same" (Yates 58n1).

36. In spite of the Lady's central role in the play, the Tyrant has more than twice as many lines as she, Govianus almost three times as many. (A rough count gives them 159, 387 and 456 lines respectively). In her suicide the Lady defines herself as speechless ("I have prepared myself for rest and silence / And took my leave of words" [3.1.133–34]); the Tyrant's violation forces her to speak again, but Govianus piously restores her to "the house of peace from whence she came / As queen of silence" (5.2.204–5).

37. Richard Levin credits the Lady with this conversion ("Double Plot" 221), but in fact, as in *The Revenger's Tragedy* (4.4), the virtuous woman is only a passive exemplum; the actual conversion is the work of her male guardian(s).

38. As noted above (n. 30), the Protestant hagiographer John Foxe recounts the story of Sophronia with apparent approval for her suicide (1: 245). Foxe's account—like that of Eusebius—follows shortly after general praise for the courage of the female martyrs: "Neither were the women anything at all behind [the men]; for they, being enticed to the filthy use of their bodies, rather suffered banishment, or willingly killed themselves" (1: 244; paraphrasing Eusebius bk. 8, ch. 9). Our sense of the Lady as a triumphant martyr is reinforced by Govianus's praise when he discovers her body. He casts her ordeal in traditional hagiographic terms as an athletic contest in which she is victorious:

Why, it was more
Than I was able to perform myself
With all the courage that I could take to me.
It tired me; I was fain to fall, and rest.
And hast thou, valiant woman, overcome
Thy honor's enemies with thine own white hand,
Where virgin-victory sits, all without help?
Eternal praise go with thee!—Spare not now;
Make all the haste you can. (171–79)

The Kistners, however, read the Lady's active role in this scene as a "distortion of natural authority by feminine leadership," and the Tyrant's continued pursuit of her after death as proof that her suicide is wrong (48).

39. The parallel with Vindice and Hippolito's manipulation of the Duke's body in *The Revenger's Tragedy* (4.2.207–22; 5.1.1–131) suggests a hero who relishes wit. Govianus's ironic answers to the demands of the soldiers ("What should she be?—Now I remember her. / O, she was a worthy creature / Before destruction grew so inward with her" [196–98]) sustain this effect.

40. Examples of this stock response include the Duke's murder of uncooperative beauties in *The Revenger's Tragedy;* the threats leveled against Miranda by her incestuous father in *The Two Noble Ladies* (2.2.553–55); and, in *The Martyred Soldier,* Henericke's enraged command to destroy the body of Victoria (249).

41. A tyrant's necrophilia does, however, appear briefly in *Periander,* an English play presented by the students of St. John's College, Oxford on February 13, 1608. (See the introduction to *The Christmas Prince,* xvi–xviii.) Based on classical sources, the play dramatizes the unsavory career of the eponymous Greek tyrant.

42. The blows not only echo the soldiers' knocking in act 3 (see stage directions at 139, 146, 163–64, 184), they suggest the rhythm of sexual penetration. Cf. the beating of Dorothea in *The Virgin Martyr* (4.2).

43. Lancashire observes that the song is "an ironic aubade" (*Second Maiden's* 219). Since both plays belonged to the King's Men, *Cymbeline's* influence again seems likely. Cf. Heywood's ironic use of an aubade at the parallel moment (227).

44. Lancashire points out that the black costume emphasizes the parallels between this scene and the first scene of the play (*Second Maiden's* 244).

45. For the date of *Valentinian* see Robert K. Turner in *The Dramatic Works in the Beaumont and Fletcher Canon* 4: 263, 389–90. All citations of the text are from Turner's edition. Like *The Second Maiden's Tragedy, Cymbeline* and *The Revenger's Tragedy, Valentinian* was written for the King's Men.

46. Fletcher could have read Procopius's Greek history of the Vandalic wars in Latin, Italian or French. For the relevant passages see the Loeb edition, 2: 38–47 (*History of the Wars* bk. 3, ch.4). Fletcher's second source was d'Urfé's highly romantic treatment of the story (*Astrée* pt. 2, ch. 12; 2: 492–553).

47. In Shakespeare's poem Tarquin creeps away like a "thievish dog" (736) after the rape, leaving Lucrece to rage in solitude. Heywood's Tarquin—a callous lover—cries, "I must be gone, *Lucrece* a kisse and part," before she *"flings from him"* (226).

48. In Golding's translation Philomela cries:

why doest thou from murdering me refraine?
Would God thou had it done before this wicked rape. From hence
Then should my soule most blessedly have gone without offence.
But if the Gods doe see this deede, and if the Gods I say
Be ought, and in this wicked worlde beare any kinde of sway,
And if with me all other things decay not, sure the day
Will come that for this wickednesse full dearly thou shalt pay.
Yea I my selfe rejecting shame thy doings will bewray.

And if I may have the power to come abrode, them blase I will.
In open face of all the world: or if thou keepe me still
As prisoner in these woods, my voyce the verie woods shall fill,
And make the stones to understand. Let Heaven to this give eare
And all the Gods and powers therein if any God be there.
(6.687–99)

49. Fletcher echoes Shakespeare: "And now this pale swan in her wat'ry nest / Begins the sade dirge of her certain ending" (*Lucrece* 1611–12).

50. Like Philomela, d'Urfé's Isidore becomes monstrous in her hatred:

> *elle luy portoit tant de haine, qu'elle ne le peust croire mort avant que l'avoir veu. Elle sort donc de son logis, s'en va droit au pallais, et voyant le corps sans teste, se lave les mains de son sang, et receut un si grand contentement de sa mort, que la joye luy dissipant entierement les forces et les esprits, elle tomba morte de l'autre costé.* (2: 538)

51. There seems to me no doubt that we are to condemn the decision to take revenge on Valentinian, as well as to betray Aecius. Aecius teaches, by precept and example, the superior virtue of loyal patience (cf. especially his sermon at 1.3.17–31, 79–98; his spectacular demonstration of patience at 1.3.121–241; and his "conversion" of Pontius from sedition at 4.4.182–209). This reading is supported by d'Urfé, where the narrator explicitly denounces Maximus for his treason (2: 538). Bowers also reads the revenge as evil (151); but Gossett argues that Maximus is wrong only in pursuing an indirect route to it (308–9).

52. According to Procopius, Aetius had in fact just conquered Attila (Loeb 2: 41).

53. Lovelace's glowing tribute to Fletcher—published in *Lucasta* (1649)—testifies to the contemporary impact of this death. Pausing in his general eulogy of Fletcher's "Bright Spirit," Lovelace praises *Valentinian* in particular—the only play he mentions by name:

> And now thy purple-robed *Tragedy*,
> In her imbroider'd Buskins, cals mine eye,
> Where the brave *Aetius* we see betray'ed,
> T'obey his Death, whom thousand live obey'd. (66)

54. Cf. the homoerotic fight to the death between Brutus and Sextus in Heywood's *The Rape of Lucrece* (252).

55. Fletcher echoes Govianus on the Tyrant's death ("Well, he's gone, / And all the kingdom's evils perish with him." [5.2.193–94]); and Antonio surveying the carnage at the end of *The Revenger's Tragedy* ("Bear up / Those tragic bodies; 'tis a heavy season. / Pray heaven their blood may wash away all treason" [5.3.127–29]).

56. Rowley probably wrote *All's Lost By Lust* for Prince Charles's Men at the Phoenix (Bentley 5: 1018–20). Citations of the text are from Edgar C. Morris's Belles Lettres edition.

57. See above, pp. 9–10 and 167n33.

58. See also Stork's introduction to his edition of the play, 69.

59. Orlin, who observes that this legend was widely known in England, posits an association between the "Roderigo of legend, who was known to have betrayed his homeland to the Moors for the sake of illicit desire, and the Roderigo of *Othello*, who vows to 'sell all my land' (1.3.377–80) to pursue by proxy his adulterous courtship of Desdemona" (205).

60. *La malheureuse devait porter désormais la responsabilité des maux qui s'abattirent sur l'Espagne, du jour où ce pays tomba aux mains des Musulmans. Toute une littérature allait être inspirée par la fille du comte Julien: de nombreux récits de date très postérieure et des poèmes du Romancero rapportent comment, en se baignant dans le Leveled, à Tolède, elle fut aperçue par Roderic: ils l'affublent du nom de Florinda et du sobriquet infamant de "Caba" ou "Cava" (du mot arabe qui signifie "prostituée").* (1: 14–15)

Several of the romances appear in Spanish in Barnstone's anthology (201–5).

61. Burt points out that the romances dealing with Roderick's loss of Spain are united by a "Fall of Man" motif, in which "la Cava" plays the role of Eve (435–39).

62. Rowley spells the king's name "Rodericke" and "Rodoricke."

63. Rowley may have read the story of the rape in Thomas Milles's *Treasurie of Auncient and Moderne Times* (684–85), although neither Stork in his introduction to the play (69–70) nor Bentley (5: 1020) suggest this. Milles does not include the story of the enchanted tower, however, so Rowley must have known at least one other version of Roderick's fall. His apparent sensitivity to the Spanish meaning of "la Cava" suggests that he was not limited to English sources. He may have named Jacinta after one of the victims of the lustful *Comendador* in Lope de Vega's *Fuente Ovejuna* (2.11), first printed in 1619. Like Rodericke's, the *Comendador*'s sexual crimes provoke a rebellion.

64. Cf. Valentinian's pander on his failure to move Lucina: "She pointed to a Lucrece, that hung by, / And with an angry looke, that from her eyes / Shot Vestall fire against me, she departed" (1.2.92–94).

65. Both the Lucreces and Lucina are merely pathetic before they are raped. The nearest approach to a curse is Virginia's prophecy: "Remember yet the Gods, O *Appius*, / Who have no part in this. Thy violent Lust/ Shall like the biting of the invenom'd Aspick, / Steal thee to hell" (4.1.256–59).

66. Ovid's Philomela too is self-destructive: When she "sawe the sworde she hoapt she should have dide, / And for the same hir naked throte she gladly did provide" (*Metamorphoses* 6.705–6).

67. Woodbridge, however, reads Jacinta's anger as part of a literary trend in this decade to celebrate "assertive women" (*Women* 244).

68. Lamb's observation is supported by the central role of anger in the early modern representations of witchcraft: See above, pp. 16 and 170n52 and Dolan 196–98. John Stearne, in *A Confirmation and Discovery of Witchcraft*

(1648), declares that women are "commonly impatient, and being displeased more malicious, and so more apt to revenge according to their power, and thereby fit instruments for the Deville" (11; qtd. MacFarlane *Witchcraft* 161). Findlay develops the link between the demonic female and revenge in her discussion of revenge tragedies (49–86).

69. This interpretation is reinforced by the parallel action of the subplot. Blinded by lust, a nobleman marries beneath his station (1.3). When his wife finds out that he has married again, she attempts to kill him, but by mistake has his friend murdered instead (4.2). (She uses her Moorish servant to strangle him, and thus a Moor features as the instrument of revenge in both plots.) The wife's soliloquy at 4.2.47–55 strongly suggests dementia.

70. Rowley's substitution of pander for king allows Jacinta to vent her anger without any restraint. In Lucina's post-rape confrontation with the emperor she upbraids him bitterly but never loses consciousness of his rank (*Valentinian* 3.1.35–97).

71. Thus, for example, Shakespeare's witches work with the "poisoned entrails" of a venomous toad, as well as "fillet of fenny snake," "eye of newt, and toe of frog" (*Macbeth* 4.1.5, 12, 14).

72. A similar pattern of corruption emerges clearly in Middleton's two late tragedies, *The Changeling* and *Women Beware Women*. In the latter Bianca uses the language of disease to describe her rape, figuring it as a contaminating penetration of her vision that leaves her "leprous." (See above, p. 184n54). We see the "poison" at work immediately as she, like Jacinta, venomously curses the panders, Guardiano and Livia. Bianca's consequent disaffection from her husband and their home signals her moral corruption. (Her mother-in-law compares her to "the new disease" [3.1.13], and a "plague" [3.2.90–91]). Flagrant adultery, cruelty, hypocrisy, attempted murder and finally suicide complete Bianca's decline. In *The Changeling* Beatrice displays "a kind of whoredom" in her heart (3.4.143–44) and is punished by a sexual coercion that confirms her corruption.

73. The image of the poisoned vault picks up a metaphor used by Jacinta before the rape: When the king tells her he has "a burning feaver," which she must "lay balsum to," she replies, "Poyson be it, / A serpentine and deadly aconite" (2.1.127–30), effectively wishing her body a source of poison.

74. Cf. Prospero's warning to Ferdinand on his betrothal:

> take my daughter: but
> If thou dost break her virgin-knot before
> All sanctimonious ceremonies may
> With full and holy rite be minister'd,
> No sweet aspersion shall the heavens let fall
> To make this contract grow; but barren hate,
> Sour-eye'd disdain and discord shall bestrew
> The union of your bed with weeds so loathly
> That you shall hate it both. (*The Tempest* 4.1.14–22)

75. Again a play on the Spanish *cavar* may lie behind Rodericke's words. In addition to its primary meaning of "to dig," the verb carries a metaphorical sense of cognitive penetration (*Diccionario de la Lengua Española*, 1956: *cavar* 3). Burt observes that in the Spanish romances the king's entry into the forbidden temple was associated with the taste of the forbidden fruit in Eden (435–36).

76. Rodericke's invasion of "hell" also points to the bawdy pun implicit in the play's use of the virgin/castle analogy: "Hell" carried a slang sense of "vagina" (cf. Shakespeare's Sonnet 144).

77. Finke and Shichtman observe that "the demonization of those 'others' who populate contested territories [. . .] almost always includes a sexual component" (70–71).

78. Rowley's presentation of the Moor seems unambiguous to me. Woodbridge, however, believes that after his succession to the throne "everyone will live happily ever after" (*Women* 253).

79. The play was not printed until the Quarto of 1661, when it appeared as *The Mayor of Quinborough*—a name it retained until 1938 when Bald restored the title of the Lambarde manuscript (Bentley 4: 884; *Hengist* xxiv). The Quarto also omitted 175 lines of the original text, perhaps first cut in the 1630s at a time of heightened censorship (Bald xxxi-iii). Although the play belonged to the King's Men in 1641, Middleton probably wrote it for another company: Bald suggests Lady Elizabeth's Men (*Hengist* xxi), Heinemann favors Prince Charles's (148). For the play's date see *Hengist*, xiii-xviii. All citations of the text are from Bald's edition.

80. Though the story of Vortiger and Hengist appears in many medieval and Renaissance histories, Middleton's main sources were the *Chronicles* of Holinshed and Fabyan (Bentley 4: 887). Bald suggests he may have also have used Geoffrey of Monmouth (*Hengist* 135–36). See *Hengist*, xxxvii-xlii, for a discussion of the sources, and the Appendix (127–36) for the early development of the Hengist legend.

81. Julia Briggs calls attention to the political-sexual analogy (483).

82. Middleton follows his sources in this respect: See Geoffrey of Monmouth 125; Holinshed 1: 556–58 and 5: 144–46; and Fabyan 61.

83. Lancashire ("Emblematic Castle" 232n45) points out that Roxena's cup of gold also connects her with the Whore of Babylon (Rev 17: 4).

84. The villains make several references to fortune(s) at critical moments: Hengist at 2.2.39–41, 2.3.26, 49, 131–33, 240, 3.3.1 and 4.3.2; Horsus at 3.1.77, and 4.3.153; Vortiger 1.1.9, 2.2.33, 3.2.28 ("Now fortune and I am sped," as he executes the rape of Castiza), and 3.3.297–98; and Roxena at 3.1.8, and 47. When Constantius refuses the crown, he says, "No such mark of fortune / Comes nere my head" (1.1.66–67). The operation of fortune in the play is also connected to repeated images of building, rising and falling (see Lancashire, "Emblematic Castle" 231–35, and J. Briggs 482–83).

85. For the moral significance of this castle see Lancashire's "Emblematic Castle," especially 231–34.

86. Heinemann makes this point.
87. *Hengist* echoes *Titus Andronicus* in many respects: In both plays the tyrant's infatuation with a predatory barbarian (Goth/Saxon) noblewoman prompts his rejection of his first "love" (Lavinia/Castiza), who had been given him by her father without her consent. In both cases the foreign woman brings with her a Machiavellian lover (Aaron/Horsus), who plots—but does not execute—the rape of the rejected woman. In both cases the victim is stalked and attacked by two men in a sylvan setting. There is also a historical parallel between the Goths' presence in Rome and the Saxons' in Britain: Both plays dramatize a moment early in the wave of invasions, the beginning of the end of a community's independence.
88. The understanding of rape premised by the plot—a man's sexual intercourse with an unconsenting woman *to whom he is not married*—was enshrined in English common law by Sir Matthew Hale: "[T]he husband cannot be guilty of a rape committed by himself upon his lawful wife, for by their mutual matrimonial consent and contract the wife hath given up herself in this kind unto her husband, which she cannot retract" (1: 629). Hale's legal fiction masks the reality of property rights in the traditional definition of rape: See Clark and Lewis 161. The marital exemption from prosecution for rape was only recently abolished in Canada (1983; D. Watt 231–32), Britain (1991; Barton 54), and the United States (1993; R. Hall 8). Nevertheless, men convicted of raping their wives generally receive shorter prison sentences than do "stranger rapists" (McCormick, Seto and Barbaree).
89. This is clear in Bald's edition, based on the Lambarde ms. In the Quarto text, however, the play ends abruptly before Castiza's liberation.
90. Fletcher probably wrote *Bonduca* for the King's Men near 1612, around the same time as *Valentinian*. All citations of the play are from Cyrus Hoy's edition in *The Dramatic Works in the Beaumont and Fletcher Canon* 4: 150–259.
91. *The Historie of England,* bk. 4, ch. 10; *Holinshed's Chronicles* 1: 495. Fletcher's other sources are Dio Cassius's *Roman History,* bk. 62, 1–12 (vol. 8 in the Loeb edition, 82–105) and Tacitus's *Annals,* bk. 12, 33–37 and bk. 14, 29–39 (Loeb edition, vol. 4, 360–67, and vol. 5, 152–73).
92. Mikalachki situates Fletcher's play in her discussion of nationality and gender in early modern England (96–114), linking it to "English anxiety about the female savagery of native origins" (96).
93. As Paul Green points out, Jonson, Spenser, Heywood and Burton all represent the Queen favourably (307, 309). Mikalachki observes that "in contrast to the historiographical condemnation of Boadicea, early modern literary works were more equivocal, allowing at least a provisional place for the ancient queen in nationalist celebrations" (122).
94. Fletcher makes it clear that the loss is Bonduca's fault: 3.5.127–39; and 5.1.3–16. Paul Green (309–14) and Sandra Clark (85–88) comment astutely on the play's treatment of gender.
95. The younger daughter signs a letter as "Bonvica," however (3.2.35).

96. Paul Green remarks: "[I]n the play, Roman culpability [for the rapes] is technically non-existent; the burden of guilt is placed on the women, and in a modern sense they are viewed as criminals rather than victims" (311).

97. Mikalachki observes that a "masculine embrace" becomes "the dominant trope" in the final scenes of early modern plays about ancient Britain; a trope "invoked as a metaphor of empire, and embodied in the staged embraces of male Britons by Roman commanders and the symbolic merging of their national emblems" (104). She attributes the attraction of this "Roman embrace" to a "fear of originary feminine savagery" (114).

Chapter 5: Redeeming the Rapist

1. Walker also comments on the way a strong cultural prejudice against female physical strength affected testimony. Since only "a discordant, disorderly, dishonourable woman used violence"(9), declarations of rape had "to deflect notions of female complicity and disorderliness that could undermine them. Consequently [. . .] physical resistance was underplayed, sometimes in ways which verbally denied agency to the raped woman or girl [. . .]" (18). The cultural prejudice against female speech also operated against the credibility of a complainant. As Carolyn Williams observes, "the more modest the victim (and therefore, the more potentially credible), the harder she should find it to state her case" (106).

2. See above, pp. 5–6.

3. The anonymous *Dick of Devonshire* (c. 1626) continues this pattern: A callous young man rapes his betrothed but denies the crime in court. A pious fraud practiced by his father and the presiding judge prompts the rapist to repent; the victim's willingness to forgive and marry him secures the happy ending. By contrast, in Middleton's *Women Beware Women* (c. 1621) the heroine leaves her husband to live with the tyrant who has raped her and marries him in the last act. Here the rape is the means of the heroine's corruption rather than the villain's redemption, and the wedding is a blasphemy that precipitates their end. Nevertheless, the motif of the rapist's repentance is parodied in 4.1 when, in response to the Cardinal's stern warning, the Duke vows "never to know her [Bianca] as a strumpet more" (269) and immediately resolves to murder her husband, so they can be lawfully married (270–74).

4. All citations of *Measure for Measure* are from J. W. Lever's Arden edition. Lever believes the play was probably first performed during the summer of 1604 (xxxv).

5. See below, n. 14.

6. See Lascelles 6–8 and J. H. Smith. Corré discusses Middle Eastern analogues.

7. Shakespeare may have read Cinthio's novella (*Hecatommithi* Part 2; 8.5) in the original Italian or a French translation. He may also have known Cinthio's later dramatic version of the story, *Epitia* (1583) (summarized with

extracts in Bullough 2: 430–42). For a discussion of Shakespeare's sources see Lascelles 6–42, Bullough 2: 399–417, Lever's introduction to the Arden xxxv-xliv and Eccles in the New Variorum edition 301–5.

8. Citations of Cinthio are from the translation by Eccles in his edition of *Measure for Measure* (378–87). Epitia also argues: "[T]hat wise men were of opinion that adultery committed through force of love, and not to wrong the lady's husband, deserved lesser punishment than that which did such wrong, and that the same should be said in the case of her brother, who had done that for which he was condemned not to wrong a husband but driven by ardent love [. . .]" (379). Epitia's argument not only underlines the masculine bias of jurisprudence, it points to the phenomenon of rape as an expression of hostility to the victim's husband—an aspect of the crime treated in *The Queen of Corinth.*

9. This is in part the perspective of Richardson's Clarissa after her rape. She astonishes everyone—especially Lovelace himself—by refusing his offer of marriage. As she puts it: "I cannot consent to *sanctify* [. . .] Mr Lovelace's repeated breaches of all moral sanctions, and hazard my *future* happiness by a union with a man, through whose premeditated injuries [. . .] I have forfeited my *temporal* hopes" (Letter 369; 1141). Clarissa nevertheless forgives him as a Christian, refrains from legal prosecution, and abjures private revenge.

10. Cinthio makes the parallel between the crimes of the prisoner and the magistrate clear: "Overcome by Epitia's grace in speaking and her rare beauty, and struck by lustful appetite, he [Juriste] turned his mind to committing against her the same fault for which he had condemned Vico to death" (380).

11. Citations of *Promos and Cassandra* are from the text reprinted by Eccles in the New Variorum edition of *Measure for Measure* (305–69). References are to act, scene and page number.

12. In her psychoanalytic reading of the play, Adelman observes: "Angelo needs the Duke as the heavenly father who comes back to judge him, condemning him to death for his manifold infractions against a sacred space and then forgiving him by demonstrating that the space has not been violated after all [. . .]" (98).

13. Like Isabella and Mariana, I accept the Duke's word for the act's legality (4.1.73). (See Schanzer, Nuttall "*Measure for Measure:* the Bed-trick," and Wentersdorf for discussion.) It thus has the same kind of ambiguity as the rape in *Hengist, King of Kent:* There the rapist is the victim's husband in disguise, here the victim is the rapist's wife in disguise. In both cases one party knows the act to be legal, and one experiences it as rape. In Shakespeare, however, the deceit allows a wife to win her husband, in Middleton it allows a husband to divorce his wife.

14. Mariana also "dies" sexually for Angelo. Gless connects sexual death with redemptive suffering in his discussion of Isabella: The sacrifice demanded of her, he notes, "*could,* in its utter self-abnegation, constitute a shockingly unconventional *imitatio Christi.* (She is being asked, in a sexual sense, to 'die'

in order to 'redeem' Claudio from the 'law.')" (126). Mariana makes the sacrifice Isabella will not.

15. The Duke also requires Isabella's participation in this ordeal: She has to charge Angelo publicly and falsely with rape, and she does it reluctantly ("To speak so indirectly I am loth, / I would say the truth [. . .]" [4.6.1–2]).

16. Gorfain relates Mariana's redemptive role in this scene to the folktale tradition of "neck riddles"—riddles that save the prisoner's life by baffling the judge. In such tales "notions of what is lawful and unlawful are redefined in the redemptive action of the neck riddle, frequently an action performed by a sacrificial female who must undergo some kind of seeming vice to restore order" (116).

17. Mariana petitions the Duke in vain four times (5.1.414–15, 423–24, 426, 428) before turning to Isabella, who relents after four appeals (428–30, 434, 435–39, 440).

18. Nuttall thinks Angelo never moves beyond this deathwish and that his reprieve "perpetuates his anguish" (*"Measure for Measure:* Quid Pro Quo?" 247). Edwards, however, believes that Angelo's redemption is successful and describes his penitence and "the reception of the new man into the society of the play" as "convincing and moving" (115); see also R. Miles 200.

19. 2.2.26–187and 2.4.1–169. This is roughly twice the time devoted to the process by Whetstone, who uses around 160 lines over three scenes (Part 1: 2.3, 2.5, 3.2).

20. Lupton discusses the play's relation to hagiography, 378–80.

21. Frustrated by the failure of her "sense" to pursue his, Angelo finally declares himself at 2.4.140 ("Plainly conceive, I love you"). Once she is sure Angelo is in earnest, Isabella has only a brief reply in which she tries to turn his criminal attempt against him (148–53). Angelo sweeps it aside, delivers his brutal ultimatum and leaves (153–69): Her unreflective defiance is delivered in soliloquy (176–86).

22. See Styan 87 and 201. For Lucio as Vice see Fergusson 80 and Winston's essay on the morality play elements in *Measure for Measure.*

23. Empson thinks Isabella's "coldness, even her rationality, is what has excited" Angelo (*Structure* 274). But the whole point of the scene, it seems to me, is the display of a rising passion, provocatively at odds with chaste decorum.

24. On the pairing of female chastity with silence, and sexual activity with speech, see especially Boose.

25. Thus, for example, Jardine remarks that Isabella betrays "an obsessive fear of her own sexuality in general" (192), and Vivian Thomas believes that at "one level of consciousness she fears her own desire more than she fears brutal violation" (195). Adelman also diagnoses fear of sexuality and analyzes it at some length (96–97). Northrop Frye notes Isabella's "strong father-fixation" (20), and Garber suggests that her "fanatical chastity" is related to "a regressive return to domination by the father" (42–43). Earlier critics used more conventional terms to analyse Isabella's failure. Thus Wilson Knight:

Is her fall any less than Angelo's? Deeper, I think. With whom is Isabel angry? Not only with her brother. She has feared this choice—terribly [. . .] She cannot sacrifice herself. Her sex inhibitions have been horribly shown her as they are, naked. [. . .] She knows now that it is not all saintliness, she sees her own soul and sees it as something small, frightened, despicable, too frail to dream of such a sacrifice. (93)

Similarly Evans finds Isabella's anger equivalent to Angelo's lust (197): "[H]er cold inhumanity is thawed by rage and her outburst exhibits a spirit devoid of such humane virtues as understanding, tolerance, compassion, love" (198–99). Isabella has, of course, had many defenders (e.g., R. W. Chambers 35–44, Josephine Bennett 69–71) but they seem to me to have the worst of it: None of them satisfactorily explains the violence of her anger.

26. Augustine's version of a "monstrous ransom" analogue makes this calculus clear. After the woman has been propositioned by the lecher,

knowing that not she had power over her own body, but her husband, she reported back to him that for her husband's sake she was ready to comply, provided her husband, the conjugal master of her body to whom all her chastity was owed, decided—thus disposing of a matter properly his own—that for his life's sake this should be done. He was grateful and told her to do this; not at all deeming the intercourse to be adultery because there was no lust on her part, and again her great love for her husband demanded it, at his own bidding and will. The woman went to the mansion of the rich man, and did what the lecher wished; but she gave her body only to her husband, who was asking, not as at other times, to lie with her, but to live. (Bullough 2: 419)

If the sense of sexual substitution which is explicit here—the woman gives "her body only to her husband" in the act of sex with the lecher—persists in later versions of the story, then Isabella's charge of incest (3.1.138–39) is not simply a function of her own pathology.

27. As Tillyard points out, the Duke's prominence in the second half of the play is achieved at Isabella's expense: After he steps forward (3.1.150), "the Duke takes charge and she proceeds to exchange her native ferocity for the hushed and submissive tones of a well-trained confidential secretary" (101). See Riefer for a feminist analysis of Isabella's submission.

28. See Diehl for a recent discussion that emphasizes the reforming power of the Duke in terms of Protestant theology.

29. Massey, who played the Duke opposite Juliet Stevenson's Isabella in the 1983–84 RSC production, notes the audience's involvement with the play as "an out-an-out suspense thriller" (20). Quoting the Duke's plan of action ('We shall advise this wronged maid [. . .] by this is your poor brother saved, your honour untainted, the poor Mariana advantaged, and the corrupt

deputy scaled' [3.1.250–56]), he remarks, "At the end of this speech, if we had done our work properly, there was an audible whoop of satisfaction from the audience. The mounties were coming to the rescue!" (20).

30. Massey's description of his feelings during the performance of the fifth act emphasizes this power:

> I felt, in this whole sequence right up to the capitulation, if I may call it that, of Isabella at line 443, as if I possessed some of the proportions of Mozart's Sarastro. It felt as if I was subjecting them, Isabella, Angelo, and Mariana, to some kind of ordeal by fire. The sequence as a whole [. . .] is constructed almost ritually. I felt, in any event, *a thrust and certainty about what I was doing, a determination to achieve what I wanted,* and the struggle to bring Isabella to her knees was quite literally exhausting. [. . .] I remember that when she finally sank to her knees, I gave in to an almost trance-like state. (28–29; emphasis added)

Certainly as Massey interpreted the part, the Duke's triumph in the fifth act is an exertion of patriarchal power *over* the other characters, and, in particular, over Isabella.

31. According to Hoy ("Shares" 98–100), the play was written in 1616 or 1617 by Fletcher (act 2), Field (acts 3 and 4) and Massinger (acts 1 and 5). (Hensman, however, believes that the play is "Massinger's abortive revision, undertaken in about 1626, of a Fletcher-Field collaboration" [1: 197]). All citations of the play are from the edition of Robert Kean Turner in *The Dramatic Works in the Beaumont and Fletcher Canon* 8: 1–111.

32. The Queen thus distinguishes herself from the evil mothers in the "Lucrece" plays: Tamora, who actively incites her sons to rape Lavinia; the Duchess in *The Revenger's Tragedy* and Tullia in *The Rape of Lucrece*, who excuse their sons' sexual crimes; and the Queen in *Cymbeline*, who encourages Cloten's pursuit, if not his attempted assault, of Imogen. Unlike her predecessors too, the Queen of Corinth is monarch in her own right (3.1.26–28)

33. The only "tyrant" who does not lust after the woman he assaults is Vortiger, in *Hengist, King of Kent*, who rapes Castiza to secure a divorce.

34. Massinger's presentation of the betrothal is ambiguous. Although Merione feels herself committed to Theanor (1.2.20–22), and a trio of courtiers pronounce the prince wronged by his mother (1.1.43–45), there has been no official agreement: Like Polonius and Laertes, Merione's brother suspects the vows of the prince have been "meere courtship; all his service / But practise how to entrap a credulous Lady" (1.2.24–5). At the end of the play, however, when Massinger wants to redeem Theanor, the betrothal suddenly appears more substantial. Crates then tells the Queen: "Your Majesty [. . .] / Allow'd once heretofore of such a Contract, / Which you repenting afterwards, revok'd it, / Being fully bent to match her with *Agenor*" (5.4.187–90).

35. There are two directions for the entrance of the courtiers. The first *("Enter the rest disguis'd")* comes at line 41, shortly after Theanor draws his dagger,

the second *("Enter six disguis'd, singing and dancing [. . .]")* comes 10 lines later, immediately before Merione's last words ("Wrong me no more as ye are men"). As Turner suggests (95n), the first direction should probably be conflated with the second.

36. See Ewbank [Ekeblad], "The 'Impure Art' of John Webster" for Webster's structural use of the masque. Both plays were performed by the King's Men and Burbage, who played Ferdinand, may also have taken the part of Crates.

37. The three Furies (in Greek, *Erinyes;* in Latin, *Dirae* or *Furiae*) were Alecto, Megaera and Tisiphone (Rose 84–85). Alecto plays a major role in the *Aeneid* where Virgil vividly describes her hair of snakes (7.327–29; 445–56). In his *Masque of Queens,* Jonson associates his witches closely with the Furies. Thus the Dame summons "You fiendes, and Furies, (if yet any bee / Worse then o[u]rselves)" (218–19); her hair is "knotted, and folded with vipers; In her hand, a Torch made of a dead-Mans arme, lighted; girded w[i]th a snake" (95–98).

38. Thus Virgil locates the Furies at the entrance to Acheron (*Aeneid* 4: 280). The conventional association of the Furies with Proserpina, the Queen of the underworld, is suggested by an allusion in Sidney's *Arcadia:* "Philoclea was parting, and Miso straight behind her, like Alecto following Proserpina" (Bk. 2, ch. 16; 325).

39. Claudian, in his *The Rape of Proserpine,* describes the holiday in hell when Pluto brings back his bride. As the couple go to bed,

The Furies now forgetfull of their rage,
With softest notes, their strict revenge asswage,
Huge goblets they prepare, and drinke a fill
Of wine, in which their monstrous locks they swil;
To the *Cerastes* powre carowses deep,
(Whil'st with new light still burning fresh they keepe
　　　The festivall spent Torches;) (2: 529–35)

Leonard Digges's translation of Claudian (1617) may have prompted Fletcher's use of the Proserpina story.

40. Merione's plight distantly resembles that of Serena in *The Faerie Queene,* who is tied naked to an altar, surrounded by "salvage" men, and menaced by a priest with a knife while horns and bagpipes shriek (6.8.35–46). Similar elements appear in Milton's *Mask,* where the Lady, imprisoned in the enchanted chair, is surrounded by Comus's *"rabble"* (658.s.d.) (*"a rout of monsters headed like sundry sorts of wild beasts, but otherwise like men and women, their apparel glistering"* [92.s.d.]). Merione's situation also, of course, anticipates that of Justina in *The Two Noble Ladies,* immobile in a chair, surrounded by magician and demons, and subject to a sexual threat (5.2).

41. The word connects Theanor with the younger son of the Duchess in *The Revenger's Tragedy,* who "played a rape on Lord Antonio's wife" (1.1.109), and at his trial declares his crime "a sport" (1.2.66).

42. Although this assault is not literally a gang rape, the role of Theanor's companions is so prominent that modern studies of group rape seem to me to be apposite here. In one such study, W. H. Blanchard reports:

    The leader of the male group . . . apparently precipitated and maintained the activity, despite misgivings, because of a need to fulfill the role that the other two men had assigned to him. "I was scared when it began to happen," he says. "I wanted to leave but I didn't want to say it to the other guys—you know—that I was scared." (qtd. Griffin, *Rape* [1986] 10)

    Like the man whom Blanchard describes, Theanor clearly undertakes the assault, despite misgivings, because he needs to fulfill the role Crates has assigned him and to be seen to do so by his companions. In her study of fraternity gang rape Sanday describes how initiation rituals construct "masculine subjectivity"

    by first stressing sexual differences and then representing these differences as hierarchical [. . .] The ritual inducts pledges into the brotherhood by first producing and then resolving anxiety about masculinity. [. . .] [It] resolves the anxiety by cleansing the pledge of his supposed feminine identification and promising him a lifelong position in a purified male social order. (171)

    These rituals, she observes, are motivated by the myth of "the polluting woman, and the demanding engulfing mother," from whom the young male must liberate himself (175). For further discussion of gang rape, see Brownmiller 203–212.

43. See, for example, Dante's *Inferno,* where the Furies at the gates of Dis invoke Medusa's aid in avenging Theseus's assault on Proserpina (9.34–63). McGee traces the connections between witches, Furies, Harpies and demons.

44. The scene presents a striking contrast to the army's adulation of Virginius in *Appius and Virginia* 4.2: There the soldiers' adulation fuels a rebellion, here the soldiers cheer the suppression of a rebellion.

45. Crates's objection is interesting: He implies that while more than one rape shows a man's want of proper self-control, a single rape—or "fall"—is a "natural" function of virility.

46. This scene recalls the reconciliation of Aecius and Pontius in *Valentinian* (also positioned at the end of the fourth act). Here too men embrace through blood. It also recalls the reconciliation of the blood-covered Virginius and his soldiers that concludes the fourth act of *Appius and Virginia.*

47. Beliza also implies that Merione's marriage to Theanor would never be sexually satisfying:

    Consider, foole, before it be too late,
    What joyes thou canst expect from such a husband,
    To whom thy first, and what's more, forc'd embraces,
    Which men say heighten pleasure, were distastefull. (95–98)

Merione replies primly that she did not enjoy the "embraces" of the prince because "they were unlawfull, / Unbless'd by *Hymen,* and left stings behind them, / Which from the marriage-bed are ever banish'd" (99–101). The exchange illustrates clearly the assumption implicit in many of these plays: That rape is just *illicit* sex, and a source of pleasure for the unconsenting woman— indeed, a greater source of pleasure than legal intercourse would be.

48. Pearse, however, defends the play against the charge of sensationalism, arguing that Fletcher carefully subordinates "the rape episode to the general design" (160).

49. The play was licensed on July 9, 1623 for performance by the Lady Elizabeth's Servants and published in 1653, with a title page ascribing it to Middleton and Rowley (Bentley 4: 891–92). Modern scholars, however, have been skeptical of Middleton's involvement in the play. Studies of authorship favor Ford for the main plot and Dekker for the gypsy scenes (Oliphant 2: 18; Lake 218–30). Brittin, however, finds "strong evidence" that Rowley was responsible for the gypsy subplot (97), and believes Middleton might have written some of the main plot and songs (98–99). All citations of the play are from Bullen's edition of Middleton's works (6: 13–230).

50. Burelbach distinguishes four separate plots in the play and points out that "all are variations on the theme of the prodigal son" (38).

51. Another of the *Exemplary Novels,* "La Gitanilla" ("The Little Gypsy Girl") is the source of the subplot. According to Bentley, Cervantes's *Novelas Ejemplares* were published in Madrid in 1613 and translated into French within a year. No English translation before 1623 has been discovered (4: 895).

52. The Kistners note that "moral darkness, symbolized by the darkened stage, and its opposites, truth and light, are recall'd throughout the play in images of night, blackness, sunlight, and moonlight" (14). Clara's name is itself part of the dark/light motif (cf. Clarissa Harlowe). Roderigo's name may have been chosen for Spain's famous tyrant-rapist, the villain in Rowley's *All's Lost By Lust.*

53. We also learn that "Roderigo has into his father's house / A passage through a garden" (1.2.9–10). Given the frequent use of gardens for sexual encounters (like Angelo's) in medieval and Renaissance literature, Roderigo's passage to "his father's house" through the garden suggests the passage to manhood through the garden of sexual sin.

54. In the source story Rodolfo leaves Leocadia locked in his chamber in order to ask his friends for advice; when he comes back he simply takes her away blindfolded and leaves her in the street (317–18).

55. Shakespeare's Tarquin, however, "hates himself for his offence" (738), and leaves Lucrece's bed "a heavy convertite" (743).

56. *Corregidor,* the Spanish term for mayor or chief magistrate, comes from the verb *corregir,* to correct, and is thus an appropriate title for Fernando, who undertakes the moral correction of his son. In Cervantes's story Rodolfo's father has no official position.

57. Cervantes's heroine also corroborates her story with a stolen crucifix, but, by contrast, it has no particular significance for Rodolfo: "[H]aving returned home, and having missed the crucifix, [he] guessed who had taken it, but

gave himself no concern about it. To a person of his wealth such a loss was of no importance [. . .]" (320).

58. Clara's use of blood to write her story connects her with Bel-imperia in *The Spanish Tragedy* (3.2.24–31), Tamyra in *Bussy D'Ambois* (5.1.176–78) and Annabella in *'Tis Pity She's a Whore* (5.1.31–47). (If the attribution of *The Spanish Gypsy* to Ford is correct, then the latter connection seems particularly interesting.) However, while Bel-imperia and Annabella write in blood because they are imprisoned, Clara, like Tamyra, *chooses* to write in blood. (Obviously the "bloody characters" are appropriate to her pain.) Nevertheless, all four women are "unchaste," and the blood seems to signify the punishment they suffer. All four letters are also connected to a revenge motif: Bel-imperia tries to incite revenge, Annabella and Tamyra try to prevent it and Clara establishes her claim to the revenge she forgoes.

59. The playwrights have added this dimension to the story. It picks up a theme—the conflicting demands of parental affection and justice—that goes back through *The Queen of Corinth* and *The Revenger's Tragedy* to the narrative sources of the "monstrous ransom" story in which the emperor loves the erring magistrate like a son (see J. H. Smith 253; in Cinthio's tale, Juriste is "very dear to" to the Emperor [378]). Cf. too the preference of Euarchus for justice over the lives of his son and nephew at the end of Sidney's *Arcadia* (bk. 5, ch. 8; 842).

60. Significantly, however, there is no suggestion in the play that Clara is unequal to Roderigo in wealth or rank: She is an eminently suitable match for him and thus Fernando does not really sacrifice anything in accepting her as a daughter-in-law. By contrast, the family of Cervantes's Leocadia, though of noble blood, is impoverished and powerless either to enforce a settlement or exact revenge (317).

61. Fernando's recovery of a daughter-in-law from the "unfathom'd seas of matchless sorrows" in 3.3 foreshadows the recovery of his own daughter (Constanza/Pretiosa, the eponymous Spanish gypsy) in 5.3. There Fernando's ecstatic response to the revelation of Constanza's identity (41–71) recalls Pericles's response to Marina.

62. Welles emphasizes the importance of female agency in the resolution of Cervantes's tale: "It is the bonding between women (Leocadia and Doña Estefanía [Rodolfo's mother]) that enables the plot to unfold, for they resist the cultural assumption of female passivity. By breaking the silence of shame imposed by the male honor code, they risk a resolution" (247).

Conclusion

1. The marital exemption has since been abolished: See above, ch. 4, n.88.

2. In addition to Jonson's Celia, other sexually threatened heroines of this type include the eponymous heroine of Marston's *Sophonisba* and Castabella in *The Atheist's Tragedy*. In each case the woman is saved from rape either by the power of her chastity or the providential intervention of a rescuer: Thus

when Celia cries "O, just God!," Bonario leaps out to save her (*Volpone* 3.7.265); and Castabella prays to "patient Heaven" (4.3.163) just before Charlemont springs up to rescue her (4.3.175).

3. In Massinger's *Unnatural Combat* (1624–25), Theocrine's rape transforms her physical appearance. "Ha! Who is this?" exclaims Malefort, "how alter'd! How deform'd!"

> It cannot be. And yet this creature has
> A kinde of a resemblance to my daughter,
> My *Theocrine!* But as different
> From that she was, as bodies dead are in
> Their best perfections, from what they were
> When they had life and motion.

"Tis most true sir," replies Theocrine,

> I am dead indeed to all but misery.
> O come not neere me sir, I am infectious;
> To look on me at distance is as dangerous
> As from a pinacles cloud-kissing spire,
> With giddy eyes to view the steepe descent,
> But to acknowledge me a certaine ruine. (5.2.190–202)

4. The language of pollution in these plays is echoed in the court records analyzed by Gowing, where whoredom "is visualized as polluting the honesty of women, households, and the neighbourhoods of the street and parish" ("Gender" 17).

5. Lyndal Roper, analyzing the language of disciplinary ordinances in early modern Germany, observes that women's bodies "were thought to have weak boundaries in a sexual sense. Sexually permeable, their wombs were constantly alive, and open to male invasion. [. . .] Female bodies [. . .] could bring pollution on society through their sexual openness" (53).

6. Cf. Woodbridge on Lavinia and Lucrece: "Like smashing an atom, smashing these women releases tremendous power. When bodily violence propels them into dangerous margins, when ceasing to be chaste wives they transgress confining boundaries, they release an uncanny power upon the world" (*Scythe* 80).

Works Cited

Abd al-Hakam, Ibn. *The History of the Conquest of Spain.* Trans. John Harris Jones. New York: Franklin, 1969

Acts of the Christian Martyrs. Trans. H. Musurillo. Oxford: Clarendon P, 1972.

Adams, Henry Hitch. *English Domestic or Homiletic Tragedy 1575 to 1642.* New York: Columbia UP, 1943.

Adelman, Janet. *Suffocating Mothers.* New York: Routledge, 1992.

Adler, Doris. *Philip Massinger.* Boston: Twayne, 1987.

Aelfric. *Lives of the Saints.* 3 vols. Ed. Walter W. Skeat. Early English Text Society os 76, 1881; os 82, 1885; os 94, 1890; os 114, 1900. London: Oxford UP, 1966.

Akrigg, G. P. V. *Jacobean Pageant.* 1962. New York: Athenaeum, 1967.

Aldhelm, St. *Opera.* Ed. R. Ehwald. Monumenta Germanica Historica: Auctorum Antiquissimorum 15. Berlin: Weidmann, 1919.

———. *The Poetic Works.* Trans. M. Lapidge and J. L. Rosier. Cambridge: Brewer, 1985.

———. *The Prose Works.* Trans. M. Lapidge and M. Herren. Ipswich: Brewer, 1979.

Allison, A. F. and D. M. Rogers. "A Catalogue of Catholic Books in English Printed Abroad or Secretly in England 1558–1640." 2 parts. *Bibliographical Studies* 3.3–4 (1956).

Ambrose, St. *Opera. Patrilogae Latinae* 16. Paris, 1880.

———. *Select Works and Letters.* Trans. H. De Romestin. Nicene and Post-Nicene Fathers, 2nd ser., vol. 10. New York, 1896.

Amussen, Susan D. "Gender, Family and the Social Order, 1560–1725." Fletcher and Stevenson 196–217.

Anderson, Bonnie S. and Judith P. Zinsser. *A History of their Own.* Vol. 1. New York: Harper, 1988.

Annals of English Drama 975–1700. Alfred Harbage. Revised by S. Schoenbaum. London: Methuen, 1964. *Supplement.* S. Schoenbaum. Evanston: Northwestern U, 1966. *Second Supplement.* S. Schoenbaum. Evanston: Northwestern U, 1970.

Atwood, Margaret. *The Handmaid's Tale.* Toronto: McLelland and Stewart, 1985.

Augustine, St. *Concerning the City of God Against the Pagans.* Trans. H. Bettenson. Harmondsworth: Penguin, [1972].

B., R. *Apius and Virginia. Tudor Interludes.* Ed. Peter Happé. Harmondsworth: Penguin, 1972. 271–317.

Bach, Rebecca Ann. "The Homosocial Imaginary of *A Woman Killed With Kindness*." *Textual Practice* 12 (1998): 503–24.

Baines, Barbara. "Effacing Rape in Early Modern Representation." *English Literary History* 65 (1998): 69–98.

Baring-Gould, S. *The Lives of the Saints*. 16 vols. Rev. ed. Edinburgh: Grant, 1914.

Barnstone, Willis, ed. *Spanish Poetry From Its Beginnings Through the Nineteenth Century.* New York: Oxford UP, 1970.

Barton, Chris. "Marital Rape—It is." *Law Teacher* 26 (1992): 54–57.

Barry, Jonathan, Marianne Hester and Gareth Roberts, eds. *Witchcraft in Early Modern Europe.* Past and Present Publications. Cambridge: Cambridge UP, 1996.

Bashar, Nazife. "Rape in England between 1550 and 1700." The London Feminist History Group 28–42.

Beattie, J. M. *Crime and the Courts in England 1660–1800*. Princeton: Princeton UP, 1986.

Beaumont, Francis and John Fletcher. *Philaster.* Ed. Andrew Gurr. Revels Plays. London: Methuen, 1969.

Beneke, Timothy. *Men on Rape.* New York: St. Martin's P, 1982.

Bennett, Josephine. *Measure for Measure as Royal Entertainment.* New York: Columbia UP, 1966.

Bennett, Judith. "Medieval Women, Modern Women: Across the Great Divide." *Culture and History 1350–1600.* Ed. David Aers. Detroit: Wayne State UP, 1992. 147–75.

Benson, Pamela Joseph. *The Invention of the Renaissance Woman.* University Park, PA: Pennsylvania State UP, 1992.

Bentley, G. E. *The Jacobean and Caroline Stage.* 7 vols. Oxford: Clarendon P, 1941–56.

Bergeron, David M. "Sexuality in *Cymbeline*." *Essays in Literature* 10 (1983): 159–68.

Berry, Philippa. "Women, Language and History in *The Rape of Lucrece*." *Shakespeare Survey* 44 (1991): 33–39.

Bibliotheca Sanctorum. 14 vols. Rome: Instituto Giovanni XXIII, 1961–87.

Bindoff, S. T., J. Hurstfield and C. H. Williams, eds. *Elizabethan Government and Society.* London: U of London, Athlone P, 1961.

Blackstone, Sir William. *The Commentaries on the Laws of England.* Ed. Robert Malcolm Kerr. 4 vols. London, 1876.

Blench, J. W. *Preaching in England in the Late Fifteenth and Sixteenth Centuries.* Oxford: Blackwell, 1964.

The Boke of Justices of Peas. 1506. Amsterdam: Theatrum, 1976.

The Book of Common Prayer, 1559. Ed. J. E. Booty. Washington, DC: Folger, 1976.

Boose, Lynda E. "Scolding Brides and Bridling Scolds: Taming the Woman's Unruly Member." *Shakespeare Quarterly* 42 (1991): 179–213.

Bornstein, Diane. *The Lady in the Tower.* Archon Books. Hamden: Shoe String, 1983.

Bossy, John. "The English Catholic Community 1603–1625." *The Reign of James VI and I.* Ed. Alan G. R. Smith. London: MacMillan, 1973. 91–105.

Bowers, Fredson Thayer. *Elizabethan Revenge Tragedy.* Princeton: Princeton UP, 1940.

Bradbrook, M. C. *The Living Monument.* Cambridge: Cambridge UP, 1976.

Breitenberg, Mark. *Anxious Masculinity in Early Modern England.* Cambridge Studies in Renaissance Literature and Culture 10. Cambridge: Cambridge UP, 1996.

Bridenbaugh, Carl. *Vexed and Troubled Englishmen.* New York: Oxford UP, 1968.

Briggs, Julia. "Middleton's Forgotten Tragedy: *Hengist, King of Kent.*" *Review of English Studies* ns 41 (1990): 479–95.

Briggs, Robin. "'Many Reasons Why'": Witchcraft and the Problem of Multiple Explanation." Barry, Hester and Roberts 64–95.

Brittin, Norma. *Thomas Middleton.* New York: Twayne, 1972.

Brown, Laura. *Ends of Empire: Women and Ideology in Early Eighteenth-Century English Literature.* Ithaca: Cornell UP, 1993.

Brown, Peter. *The Body and Society.* New York: Columbia UP, 1988.

Brownlow, F. W. "John Shakespeare's Recusancy: New Light on an Old Document." *Shakespeare Quarterly* 40 (1989): 186–91.

Brownmiller, Susan. *Against Our Will.* 1975. New York: Bantam, 1976.

Brundage, James A. *Law, Sex, and Christian Society in Medieval Europe.* Chicago: U of Chicago P, 1987.

Bruster, Douglas. *Drama and the Market in the Age of Shakespeare.* Cambridge Studies in Renaissance Literature and Culture 1. Cambridge: Cambridge UP, 1992.

Bugge, John. *Virginitas.* International Archives of the History of Ideas 17. The Hague: Nijhoff, 1975.

Bullough, Geoffrey, ed. *Narrative and Dramatic Sources of Shakespeare.* 8 vols. London: Routledge, 1957–75.

Burelbach, Frederick M., Jr. "Theme and Structure in *The Spanish Gypsy.*" *The Humanities Association Bulletin* 11.2 (1968): 37–41.

Burgess, Ann, ed. *Rape and Sexual Assault.* 2 vols. New York: Garland, 1985–88.

Burks, Deborah. "'I'll Want My Will Else': *The Changeling* and Women's Complicity with Their Rapists." *English Literary History* 62 (1995): 759–90.

Burt, John R. "The Motif of the Fall of Man in the '*Romancero del rey Rodrigo.*'" *Hispania* 61 (1978): 435–42.

Cahn, Susan. *Industry of Devotion.* New York: Columbia UP, 1987.

Calvin, John. *Institutes of the Christian Religion.* Trans. H. Beveridge. 2 vols. London: James Clarke, n.d.

Camelot, P. T. "Virginity." *New Catholic Encyclopedia.* 16 vols. New York: McGraw Hill, 1967–79.

Carroll, Berenice A. *Liberating Women's History.* Urbana: U of Illinois P, 1976.

Carter, Angela. *The Sadeian Woman.* London: Virago, 1979.

Carter, John Marshall. *Rape in Medieval England.* Lanham: UP of America, 1985.

Cary, Elizabeth, Lady Falkland. *The Tragedy of Mariam.* Ed. Barry Weller and Margaret W. Ferguson. Berkeley: U of California P, 1994.

Cervantes Saavedra, Miguel de. *Exemplary Novels.* Trans. Walter K. Kelly. London: Bell, 1908.

Chamberlain, John. *The Letters of John Chamberlain.* Ed. Norman Egbert McLure. 2 vols. Philadelphia: American Philosophical Society, 1939.

Chambers, E. K. *The Elizabethan Stage.* 4 vols. Oxford: Clarendon P, 1923.

Chambers, R. W. *The Jacobean Shakespeare and* Measure for Measure. 1937. Freeport: Books for Libraries P, 1970.

Champion, Larry S. *Thomas Dekker and the Traditions of English Drama.* American University Studies, Series 42: English Language and Literature 27. New York: Lang, 1985.

Chapman, George. *The Tragedy of Bussy D'Ambois.* Ed. Nicholas Brooke. Revels Plays. London: Methuen, 1964.

Charlton, Kenneth. *Education in Renaissance England.* London: Routledge, 1965.

Chaucer, Geoffrey. *Works.* Ed. W. W. Skeat. 6 vols. 2nd ed. Oxford: Clarendon P, 1899–1900.

Chaytor, Miranda. "Husband(ry): Narratives of Rape in the Seventeenth Century." *Gender and History* 7 (1995): 378–407.

The Christmas Prince. Ed. Frederick S. Boas and W. W. Greg. Malone Society Reprints 47b. Oxford: Oxford UP, 1922.

Chrysostom, John. *On the Necessity of Guarding Virginity.* Trans. E. Clark. *Jerome, Chrysostom and Friends.* Studies in Women and Religion 1. New York: Mellen, 1979.

Clark, Alice. *Working Life of Women in the Seventeenth Century.* 1919. London: Routledge, 1982.

Clark, Anna. *Women's Silence, Men's Violence.* London: Pandora, 1987.

Clark, Elizabeth A. *Ascetic Piety and Women's Faith.* New York: Mellen, 1986.

Clark, Lorenne and Debra Lewis. *Rape.* Toronto: Women's P, 1977

Clark, Sandra. *The Plays of Beaumont and Fletcher.* New York: Harvester-Wheatsheaf, 1994.

Clark, Stuart. "Inversion, Misrule and Witchcraft." *Past and Present* 87 (1980): 98–127.

Claudian. *The Rape of Proserpine.* Trans. Leonard Digges. Ed. H. H. Huxley. English Reprint Series. Liverpool: Liverpool UP, 1959.

Clyomon and Clamydes. Ed. Betty J. Littleton. Studies in English Literature 35. The Hague: Mouton, 1968.

Cockburn, J. S., ed. *Crime in England 1550–1800.* London: Methuen, 1977.

———. "The Nature and Incidence of Crime in England 1559–1625: A Preliminary Survey." Cockburn, *Crime in England* 49–71.

Coleridge, Samuel Taylor. *Coleridge on the Seventeenth Century.* Ed. Roberta Florence Brinkley. Introduction by Louis I. Bredvold. [Durham, NC]: Duke UP, 1955.

Collinson, Patrick. *The Birthpangs of Protestant England.* Houndmills: MacMillan, 1988.

The Columbia Encyclopedia. 3rd ed. 1963.

Corré, Alan D. "The Lecher, the Coward and the Virtuous Woman." *Folklore* 92 (1981): 25–29.

Couliano, Ioan P. *Eros and Magic in the Renaissance.* Trans. Margaret Cook. Foreword by Mircea Eliade. Chicago: U of Chicago P, 1987.

Crane, Mary Thomas. "Male Pregnancy and Cognitive Permeability in *Measure for Measure.*" *Shakespeare Quarterly* 49 (1998): 269–92.

Crawford, Patricia. "From the Woman's View: Pre-Industrial England, 1500–1700." *Exploring Women's Past.* Ed. Patricia Crawford, Margaret Anderson et al. Sydney: Allen, 1983.

———. *Women and Religion in England 1500–1720.* London: Routledge, 1993.

Cressy, David. *Birth, Marriage and Death.* Oxford: Oxford UP, 1997.

———. *Bonfires and Bells.* London: Weidenfield, 1989.

Curtis, T. C. "Quarter Sessions Appearances and their Background: A Seventeenth-Century Regional Study." Cockburn, *Crime in England* 135–54.

Daly, Mary. *Gyn/Ecology.* Boston: Beacon P, 1978.

Davies, Kathleen M. "Continuity and Change in Literary Advice on Marriage." *Marriage and Society.* Ed. R. B. Outhwaite. London: Europa, 1981. 58–80.

Davis, Natalie Zemon. *Society and Culture in Early Modern France.* Stanford: Stanford UP, 1975.

Dawson, Anthony B. "*Women Beware Women* and the Economy of Rape." *Studies in English Literature* 27 (1987): 303–20.

De Groot, Roger D. "The Crime of Rape *temp.* Richard I and John." *Journal of Legal History* 9 (1988): 324–34.

De Voragine, Jacobus. *The Golden Legend.* Trans. William Caxton. 7 vols. London: Dent, 1931.

Dekker, Thomas and Philip Massinger. *The Virgin Martyr. The Dramatic Works of Thomas Dekker.* Ed. F. T. Bowers. Vol. 3. Cambridge: Cambridge UP, 1958. 365–480.

Deloney, Thomas. *Novels.* Ed. Merritt E. Lawlis. Bloomington: Indiana UP, 1961.

Detmer, Emily. "'This helpless smoke of words': Locating the Agency in Withholding Consent and Reporting Rape." Unpublished paper, presented at the Twenty-Seventh Annual Meeting of the Shakespeare Association of America, San Francisco, April 1999.

———. "The Politics of Telling: Women's Consent and Accusations of Rape in English Renaissance Drama." Diss. Miami U, 1997.

Diccionario de la Lengua Española. 16th ed. Madrid: Academia Española, 1956.

Dick of Devonshire. Ed. J. G. McManaway and M. R. McManaway. Malone Society Reprints. Oxford: Oxford UP, 1955.

Dictionary of National Biography. Ed. Leslie Stephen and Sidney Lee. 63 vols. London: Smith-Elder, 1885–1900.

Diehl, Huston. "'Infinite Space': Representation and Reformation in *Measure for Measure.*" *Shakespeare Quarterly* 49 (1998): 393–410.

Digangi, Mario. "Pleasure and Danger: Measuring Female Sexuality in *Measure for Measure.*" *English Literary History* 60 (1993): 589–609.

Dinshaw, Carolyn. "Rivalry, Rape, and Manhood: Gower and Chaucer." A. Roberts 137–160.

Dio Cassius. *Dio's Roman History.* Trans. Earnest Cary. 9 vols. Loeb Classical Library. London: Heinemann, 1914.

Dionysius of Halicarnassus. *The Roman Antiquities of Dionysius of Halicarnassus.* Trans. Earnest Cary. 7 vols. London: Heinemann, 1937.

Dolan, Frances E. *Dangerous Familiars.* Ithaca: Cornell UP, 1994.

Donaldson, Ian. *The Rapes of Lucretia.* Oxford: Clarendon P, 1982.

Douglas, Mary. *Purity and Danger.* London: Routledge, 1966.

Dubinsky, Karen. *Improper Advances.* The Chicago Series on Sexuality, History, and Society. Chicago: U of Chicago P, 1993.

Dubrow, Heather. "A Mirror For Complaints." *Renaissance Genres.* Ed. Barbara Kiefer Lewalski. Harvard English Studies 14. Cambridge, MA: Harvard UP, 1986. 399–417.

Duffy, Eamon. *The Stripping of the Altars.* New Haven: Yale UP, 1992.

Dworkin, Andrea. *Right-wing Women.* New York: Pedigree, 1983.

Easton, Martha. "Saint Agatha and the Sanctification of Sexual Violence." *Studies in Iconography* 16 (1994): 83–118.

Edwards, Philip. *Shakespeare and the Confines of Art.* London: Methuen, 1968.

Eisenstein, Elizabeth. *The Printing Press as an Agent of Change.* 2 vols. 1979. Combined paperback ed. Cambridge: Cambridge UP, 1980.

Empson, William. "The Narrative Poems." *Essays on Shakespeare.* Ed. David B. Pirie. Cambridge: Cambridge UP, 1986.

———. *The Structure of Complex Words.* 1951. London: Hogarth, 1985.

Erasmus, Desiderius. *The Comparation of a Vyrgin and a Martyr.* 1537. Trans. Thomas Paynell. Introduction by W. J. Hirten. Gainesville: Scholars' Facsimiles, 1970.

Erickson, Amy Louise. *Women and Property in Early Modern England.* New York: Routledge, 1993.

Eusebius. *The Ecclesiastical History.* Trans. Kirsopp Lake. 2 vols. Loeb Classical Library. Cambridge, MA: Harvard UP, 1926.

Evans, Bertrand. *Shakespeare's Comedies.* Oxford: Oxford UP, 1960.

Ewbank [Ekeblad], Inga-Stina. "The 'Impure Art' of John Webster." *Review of English Studies* ns 9 (1958): 253–67.

———. "The Word in the Theater." *Shakespeare, Man of the Theater.* Proceedings of the Second Congress of the International Shakespeare Association, 1981. Ed. Kenneth Muir et al. Newark: Associated UP, 1983. 55–75.

Ezell, Margaret J. M. *The Patriarch's Wife.* Chapel Hill: U of North Carolina P, 1987.

Fabyan, Robert. *The New Chronicles of England and France.* London, 1811

Fawcett, Mary Laughlin. "Arms/Words/Tears: Language and the Body in *Titus Andronicus.*" *ELH* 50 (1983): 261–77.

Felperin, Howard. "Shakespeare's Miracle Play." *Shakespeare Quarterly* 18 (1967): 363–74.

Fenton, Geoffrey. *Certain Tragical Discourses of Bandello.* 1567. Ed. R. L. Douglas. 2 vols. Tudor Translations 20. London, 1898.

Fergusson, Francis. "Philosophy and Theatre in *Measure for Measure.*" Geckle 73–85.

Field, Nathan. *A Woman is a Weathercock.* Ed. A. Wilson Verity. *Nero and Other Plays.* Mermaid Series. London: Unwin, n.d. 337–412.

Findlay, Alison. *A Feminist Perspective on Renaissance Drama.* Oxford: Blackwell, 1999.

Finke, Laurie A. "Painting Women: Images of Femininity in Jacobean Tragedy." *Performing Feminisms: Feminist Critical Theory and Theatre.* Ed. Sue-Ellen Case. Baltimore: Johns Hopkins UP, 1990. 223–48.

———and Martin Shichtman. "The Mont St. Michel Giant: Sexual Violence and Imperialism in the Chronicles of Wace and Laʒamon." A. Roberts 56–74.

Fletcher, Anthony and John Stevenson, eds. *Order and Disorder in Early Modern England.* Cambridge: Cambridge UP, 1985.

Fletcher, John. *The Dramatic Works in the Beaumont and Fletcher Canon.* General ed. Fredson Bowers. 8 vols. Cambridge: Cambridge UP, 1966–92.

Ford, John. *The Broken Heart.* Ed. Brian Morris. New Mermaids. London: Benn, 1965.

———. *'Tis Pity She's A Whore.* Ed. Brian Morris. New Mermaids. London: Benn, 1968.

Fox, Robin Lane. *Pagans and Christians.* 1986. Harmondsworth: Penguin, 1988.

Foxe, John. *Acts and Monuments.* Ed. Stephen Reed Cattley. 8 vols. London, 1841.

Fraser, Antonia. *The Warrior Queens.* 1989. Harmondsworth: Penguin, 1990.

Friedman, Alice T. "Portrait of a Marriage: the Willoughby Letters of 1585–1586." *Signs* 11 (1986): 542–55.

Frye, Northrop. *The Myth of Deliverance.* Toronto: U of Toronto P, 1983.

Gajowski, Evelyn. "Lavinia as 'Blank Page': Voicelessness, Violation and Violence in *Titus Andronicus.*" Unpublished essay, 1999.

Garber, Marjorie. *Coming of Age in Shakespeare.* London: Methuen, 1981.

Gardiner, S. R. *The History of England from the Accession of James I to the Outbreak of the Civil War 1603–1642.* 10 vols. London, 1883.

Gasper, Julia. *The Dragon and the Dove.* Oxford: Clarendon P, 1990.

———. "The Sources of *The Virgin Martyr.*" *Review of English Studies* ns 42 (1991): 17–31.

Geckle, George, ed. *Twentieth Century Interpretations of* Measure for Measure. Englewood Cliffs: Prentice-Hall, 1970.

Geoffrey of Monmouth. *History of the Kings of Britain.* Trans. Sebastian Evans. Everyman's Library. London: Dent, [1963].

Gerard, John. *Autobiography of an Elizabethan.* Trans. Philip Caraman. London: Longmans, 1951.

Gesner, Carol. *Shakespeare and the Greek Romances.* Lexington: UP of Kentucky, 1970.

Girard, René. *Deceit, Desire and the Novel.* Trans. Yvonne Freccero. Baltimore: Johns Hopkins UP, 1966.

———. *Violence and the Sacred.* Trans. Patrick Gregory. Baltimore: Johns Hopkins UP, 1977.

Gless, Darryl. Measure for Measure: *The Law and the Convent.* Princeton: Princeton UP, 1979.

Goffe, Thomas. *The Courageous Turk.* Ed. Susan Gushee O'Malley. Renaissance Drama. New York: Garland, 1979.

Goolden, P. "Antiochus' Riddle in Gower and Shakespeare." *Review of English Studies* ns 6 (1955): 245–51.

Gorfain, Phyllis. "Riddling as Ritual Remedy in *Measure for Measure.*" Woodbridge and Berry 98–122.

Gossett, Suzanne. "'Best Men are Molded out of Faults': Marrying the Rapist in Jacobean Drama." *English Literary Renaissance* 14 (1983): 305–27.

Gowing, Laura. "Gender and the Language of Insults." *History Workshop Journal* 35 (1993): 1–21.

——. "Language, Power and the Law: Women's Slander Litigation in Early Modern London." Kermode and Walker 26–47.

Gramsci, Antonio. *Selections from the Prison Notebooks.* Ed. and trans. Quintin Hoare and Geoffrey Nowell Smith. New York: International, 1971.

Granville-Barker, Harley. *Prefaces to Shakespeare.* 1930. Ed. M. St. Clare Byrne. 4 vols. London: Batsford, 1963.

Gray, J. C., ed. *Mirror up to Shakespeare.* Toronto: U of Toronto P, 1984.

Greaves, R. *Society and Religion in Elizabethan England.* Minneapolis: U of Minnesota P, 1981.

Green, Douglas E. "Interpreting 'her martyr'd signs': Gender and Tragedy in *Titus Andronicus.*" *Shakespeare Quarterly* 40 (1989): 317–26.

Green, Paul D. "Theme and Structure in Fletcher's *Bonduca.*" *Studies in English Literature* 22 (1982): 305–16.

Greene, Robert. *James the Fourth.* Ed. J. A. Lavin. New Mermaids. London: Benn, 1967.

Griffin, Susan. *Rape.* 3rd ed. San Francisco: Harper, 1986.

——. "Rape: The All-American Crime." *Ramparts* Sept. 1971: 26–35.

Gurr, Andrew. *Playgoing in Shakespeare's London.* Cambridge: Cambridge UP, 1987.

——. *The Shakespearean Stage, 1574–1642.* 2nd ed. Cambridge: Cambridge UP, 1980.

Haber, Judith. "'I(t) could not choose but follow': Erotic Logic in *The Changeling.*" Unpublished paper, presented at the Twenty-Seventh Annual Meeting of the Shakespeare Association of America, San Francisco, April 1999.

Haigh, Christopher. "The Continuity of Catholicism in the English Reformation." *Past and Present* 93 (1981): 37–69.

Hale, Sir Matthew. *Historia Placitorum Coronae.* 1736. Ed. P. R. Glazebrook. 2 vols. London: Professional Books, 1971.

Hali Meidenhad. Ed. F. J. Furnivall. Early English Text Society os 18a, 1922. Rev. ed. New York: Greenwood, 1969.

Halkett, John. *Milton and the Idea of Matrimony.* New Haven: Yale UP, 1970.

Hall, Robert. *Rape in America.* Santa Barbara, CA: ABC-CLIO, 1995.

Haller, William and Malleville. "The Puritan Art of Love." *Huntington Library Quarterly* 5 (1942): 235–74.

Hanawalt, Barbara. *Crime and Conflict in English Communities, 1300–1348.* Cambridge, MA: Harvard UP, 1979.

Hannay, Margaret P. "'Faining Notable Images of Vertue': Sidney's *New Arcadia* as *Legenda Sanctorum.*" *University of Hartford Studies in Literature* 15–16.3–1 (1983–84): 80–92.

Harris, Barbara. "Marriage Sixteenth-Century Style: Elizabeth Stafford and the Third Duke of Norfolk." *Journal of Social History* 15 (1982): 371–82.

Harris, Victor. *All Coherence Gone.* Chicago: U of Chicago P, 1949.

Harris, W. V. "The Roman Father's Power of Life and Death." *Studies in Roman Law in Memory of A. Arthur Schiller.* Ed. R. S. Bagnall and W. V. Harris. Leiden: Brill, 1986.

Harvey, Susan Ashbrook. "Women in Early Syrian Christianity." *Images of Women in Antique Piety.* Ed. A. Cameron and A. Kuhrt. London: Croom Helm, 1983. 288–98.

Hastrup, K. "The Semantics of Biology: Virginity." *Defining Females.* Ed. S. Ardener. Oxford Women's Series 1. London: Croom Helm, 1978. 49–65.

Heal, Felicity. "The Idea of Hospitality in Early Modern England." *Past and Present* 102 (1984): 66–93.

Heinemann, Margot. *Puritanism and Theatre.* Past and Present Publications. Cambridge: Cambridge UP, 1980.

Helms, Lorraine. "'The High Roman Fashion': Sacrifice, Suicide, and the Shakespearean Stage." *PMLA* 107 (1992): 554–65.

Henken, Elissa R. *Traditions of Welsh Saints.* Cambridge: Brewer, 1987.

Hensman, Bertha. *The Shares of Fletcher, Field and Massinger in Twelve Plays of the Beaumont and Fletcher Canon.* 2 vols. Jacobean Drama Studies 6. Salzburg: Institut für Englische Sprache und Literatur, 1974.

Hester, Marianne. "Patriarchal Reconstruction and Witch Hunting." Barry, Hester and Roberts 288–306.

Heywood, Thomas. *Dramatic Works.* Ed. R. H. Shepherd. 6 vols. London, 1874.

———. *The Rape of Lucrece.* Ed. Allan Holaday. *Illinois Studies in Language and Literature* 34.3 (1950).

Hill, Christopher. *Society and Puritanism in Pre-Revolutionary England.* 1964. Harmondsworth: Peregrine-Penguin, 1986.

———. *The World Turned Upside Down.* 1972. Harmondsworth: Penguin, 1975.

Hogrefe, Pearl. *Women of Action in Tudor England.* Ames: Iowa State UP, 1977.

Holinshed, Raphael. *Chronicles of England, Scotland and Ireland.* 6 vols. London, 1807.

Houlbrooke, Ralph. *The English Family 1450–1700.* London: Longman, 1984.

Howard, Jean. "Scripts and/versus Playhouses: Ideological Production and the Renaissance Public Stage." *Renaissance Drama* 20 (1989): 31–49.

Hoy, Cyrus. "The Shares of Fletcher and His Collaborators in the Beaumont and Fletcher Canon (IV)." *Studies in Bibliography* 12 (1959): 91–116.

———. *Introductions, Notes and Commentaries to Texts in* The Dramatic Works of Thomas Dekker. 4 vols. Cambridge: Cambridge UP, 1980.

Hrotsvitha. *The Plays of Roswitha* [sic]. Trans. Christopher St. John. New York: Cooper Square, 1966.

Hull, Suzanne W. *Chaste, Silent & Obedient.* San Marino: Huntington, 1982.

Hunt, Mary Leland. *Thomas Dekker.* New York: Russell, 1964.

Hunter, G. K. "Sources and Meanings in *Titus Andronicus.*" Gray 171–88.

Hurstfield, Joel. "The Succession Struggle in Late Elizabethan England." Bindoff, Hurstfield and Williams 369–396.

Hutson, Lorna. *The Usurer's Daughter.* London: Routledge, 1994.

Ingram, Martin. *Church Courts, Sex and Marriage in England, 1570–1640.* Cambridge: Cambridge UP, 1987.

————. "Ridings, Rough Music and the 'Reform of Popular Culture' in Early Modern England." *Past and Present* 105 (1984): 79–113.

————. "'Scolding women cucked or washed': a Crisis in Gender Relations in Early Modern England?" Kermode and Walker 48–80.

Irwin, Joyce L. *Womanhood in Radical Protestantism 1515–1675*. Studies in Women and Religion 1. Lewiston: Wellen, 1979.

James I and VI. *Political Works*. 1918. Introduction by C. H. McIlwain. New York: Russell, 1965.

Jardine, Lisa. *Still Harping on Daughters*. Brighton: Harvester, 1983.

Jed, Stephanie H. *Chaste Thinking*. Theories of Representation and Difference. Bloomington: Indiana UP, 1989.

Jensen, Phebe. "Recusancy, Festivity and Community: The Simpsons at Gowlthwaite Hall." Unpublished paper, presented at the Twenty-Seventh Annual Meeting of the Shakespeare Association of America, San Francisco, April 1999.

Jerome, St. *Opera. Patrilogiae Latinae* 22. Paris, 1854.

Jonson, Ben. *The Masque of Queens. Ben Jonson*. Vol. 7. Ed. Herford and Simpson. Oxford: Clarendon P, 1941. 267–320.

————. *Volpone*. Ed. Brian Parker. Revels Plays. Manchester: Manchester UP, 1983.

Joplin, Patricia Klindienst. "The Voice of the Shuttle is Ours." *Stanford Literature Review* 1 (1984): 25–53.

Joshel, Sandra. "The Body Female and the Body Politic: Livy's Lucretia and Verginia." Richlin, *Pornography* 112–30.

Kahn, Coppélia. *Roman Shakespeare*. Feminist Readings of Shakespeare. London: Routledge, 1997.

Karlsen, Carol F. *The Devil in the Shape of a Woman*. 1987. New York: Vintage-Random House, 1989.

Kelly, J. N. D. *Early Christian Doctrines*. Rev. ed. San Francisco: Harper, 1960.

Kendall, Gillian Murray. "'Lend me thy hand': Metaphor and Mayhem in *Titus Andronicus*." *Shakespeare Quarterly* 40 (1989): 299–316.

Kermode, Jenny and Garthine Walker, ed. *Women, Crime and the Courts in Early Modern England*. London: UCL P, 1994.

Kirsch, Arthur. *Shakespeare and the Experience of Love*. Cambridge: Cambridge UP, 1981.

Kistner, A. L. and M. K. *Middleton's Tragic Themes*. American University Studies, Series 4: English Language and Literature 10. New York: Lang, 1984.

Kittel, Ruth. "Rape in Thirteenth-Century England: A Study of the Common Law Courts." *Women and the Law*. Vol. 2. Ed. D. Kelly Weisberg. Cambridge, MA: Schenkman, 1982. 101–15.

Klein, Joan Larsen, ed. *Daughters, Wives, and Widows*. Urbana: U of Illinois P, 1992.

Knight, G. Wilson. *The Wheel of Fire*. 1949. 4th ed. University Paperback. London: Methuen, 1960.

Kortekaas, G. A. A., ed. *Historia Apollonii Regis Tyri*. Groningen: Bouma, 1984.

Kyd, Thomas. *The Spanish Tragedy*. Ed. Philip Edwards. Revels Plays. London: Methuen, 1959.

Lake, D. J. *The Canon of Thomas Middleton's Plays*. London: Cambridge UP, 1975.

Lamb, Mary Ellen. *Gender and Authorship in the Sidney Circle.* Madison: U of Wisconsin P, 1990.

Lancashire, Anne. "The Emblematic Castle." Gray 223–41.

———. "*The Second Maiden's Tragedy:* A Jacobean Saint's Life." *Review of English Studies* ns 25 (1974): 268–79.

Larner, Christina. *Enemies of God.* Oxford: Basil Blackwell, 1981.

Lascelles, Mary. *Shakespeare's* Measure for Measure. London: U of London, Athlone P, 1953.

The Lawes Resolutions of Womens Rights. T. E. 1632. Amsterdam: Theatrum, 1976.

Lazareth, William. *Luther on the Christian Home.* Philadelphia: Muhlenberg, 1960.

Lévi-Provencal, É. *Histoire de L'Espagne Musulmane.* 3 vols. Paris: Maisonneuve, 1950.

Levin, Carole. "Advice on Women's Behavior in Three Tudor Homilies." *International Journal of Women's Studies* 6 (1983): 176–85.

———and Karen Robertson, ed. *Sexuality and Politics in Renaissance Drama.* Studies in Renaissance Literature 10. Lewiston: Edwin Mellen P, 1991.

Levin, Richard. "The Double Plot of *The Second Maiden's Tragedy.*" *Studies in English Literature* 3 (1963): 219–31.

———. "Women in the Renaissance Theatre Audience." *Shakespeare Quarterly* 40 (1989): 165–74.

The Lives of Women Saints of Our Contrie of England. Ed. C. Horstman. London: Early English Text Society, 1886.

Livy. *The Early History of Rome.* [Bks. 1–5 of *The History of Rome*]. Trans. A. de Selincourt. Introduction by R. M. Ogilvie. Harmondsworth: Penguin, 1960.

———. *The Romane Historie.* Trans. Philemon Holland. London, 1600.

The London Feminist History Group. *The Sexual Dynamics of History.* London: Pluto, 1983.

Lottes, Ilsa. "Sexual Socialization and Attitudes toward Rape." Burgess 2: 193–220.

Love, Louise. "The Saint's Play in England During the Protestant Transition." Diss. Northwestern U, 1984.

Lovelace, Richard. *Lucasta.* 1649. Menston: Scolar P, 1972.

Lupton, Julia Reinhard. "Afterlives of the Saints: Hagiography in *Measure for Measure.*" *Exemplaria* 2 (1990): 375–401.

Luther, Martin. *Works.* Ed. Jaroslav Pelikan. 54 vols. St. Louis: Concordia, 1958.

MacCaffrey, Wallace T. "Place and Patronage in Elizabethan Politics." Bindoff, Hurstfield and Williams 95–126.

McCormick, Jennifer S., Alexander C. Seto and Howard E. Barbaree. "Relationship to Victim Predicts Sentence Length in Sexual Assault Cases." *Journal of Interpersonal Violence* 13 (1998): 413–20.

MacDonald, Michael. *Mystical Bedlam.* Cambridge: Cambridge UP, 1981.

MacDonald, Robert H., ed. *The Library of Drummond of Hawthornden.* Edinburgh: Edinburgh UP, 1971.

MacFarlane, Alan. *Marriage and Love in England 1300–1840.* Oxford: Blackwell, 1986.

———. *Witchcraft in Tudor and Stuart England.* Harper Torchbook. New York: Harper, 1970.

McGee, Arthur R. "*Macbeth* and the Furies." *Shakespeare Survey* 19 (1966): 55–67.

Machiavelli, Nicolo. *The Chief Works and Others*. Trans. Allan Gilbert. 3 vols. Durham, NC: Duke UP, 1965.

Mackinnon, Catherine A. *Feminism Unmodified*. Cambridge, MA: Harvard UP, 1987.

McLaren, Angus. *Reproductive Rituals*. London: Methuen, 1984.

Maclean, Ian. *The Renaissance Notion of Women*. 1980. Cambridge: Cambridge UP, 1987.

McLuskie, Kathleen. "The Patriarchal Bard: Feminist Criticism and Shakespeare: *King Lear* and *Measure for Measure*." *Political Shakespeare*. Ed. Jonathan Dollimore and Alan Sinfield. Manchester: Manchester UP, 1985. 88–108.

———. *Renaissance Dramatists*. Atlantic Highlands, NJ: Humanities P, 1989.

McNamara, Jo Ann. "Sexual Equality and the Cult of Virginity in Early Christian Thought." *Feminist Studies* 3.3–4 (1976): 145–84.

Marcus, Leah Sinanoglou. "The Milieu of Milton's *Comus*: Judicial Reform at Ludlow and the Problem of Sexual Assault." *Criticism* 25 (1983): 293–327.

Marsden, Jean I. "Rape, Voyeurism, and the Restoration Stage." *Broken Boundaries: Women and Feminism in Restoration Drama*. Ed. Katherine M. Quinsey. Lexington: UP of Kentucky, 1996. 185–200.

Marston, John. *Sophonisba*. *The Selected Plays of John Marston*. Ed. MacDonald P. Jackson and Michael Neill. Cambridge: Cambridge UP, 1986. 395–481.

Masek, Rosemary. "Women in an Age of Transition, 1485–1714." *The Women of England from Anglo-Saxon Times to the Present*. Ed. Barbara Kanner. Hamden: Archon, 1979. 138–82.

Massey, Daniel. "The Duke in *Measure for Measure*." *Players of Shakespeare 2*. Ed. Russell Jackson and Robert Smallwood. Cambridge: Cambridge UP, 1988.

Massinger, Philip. *The Unnatural Combat*. *The Poems and Plays of Philip Massinger*. Ed. Philip Edwards and Colin Gibson. Vol. 2. Oxford: Clarendon P, 1976. 181–272.

——— and Thomas Dekker. *The Virgin Martyr*. *The Dramatic Works of Thomas Dekker*. Ed. F. T. Bowers. Vol. 3. Cambridge: Cambridge UP, 1958. 365–480.

Mattingly, Garrett. *Renaissance Diplomacy*. 1955. Harmondsworth: Peregrine-Penguin, 1965.

Middleton, Thomas. *The Changeling*. Ed. Patricia Thomson. New Mermaids. Benn: London, 1964.

———. *The Ghost of Lucrece*. Ed. J. Q. Adams. New York: Scribner's, 1937.

———. *Hengist, King of Kent*. Ed. R. C. Bald. Folger Shakespeare Library. New York: Scribner's, 1938.

———. *The Revenger's Tragedy: Attributed to Thomas Middleton*. Facsimile edition. Introduction by MacD. P. Jackson. London: Associated UP, [1983].

———. *Women Beware Women*. Ed. J. R. Mulryne. Revels Plays. London: Methuen, 1975.

———. *The Works*. Ed. A. H. Bullen. 8 vols. London, 1885–86.

————and William Rowley. *The Spanish Gipsie, and All's Lost By Lust, by Thomas Middleton and William Rowley.* Ed. Edgar C. Morris. Belles Lettres Series. Boston: Heath, 1908.

Mikalachki, Jodi. *The Legacy of Boadicea.* London: Routledge, 1998.

Miles, Margaret R. *Carnal Knowing.* Boston: Beacon P, 1989.

Miles, Rosalind. *The Problem of* Measure for Measure. London: Vision, 1976.

Milles, Thomas. *The Treasurie of Auncient and Moderne Times.* London, 1613.

Milton, John. *A Mask (Comus). Poetical Works.* Ed. Douglas Bush. London: Oxford UP, 1966. 109–39.

Miola, Robert S. "*Cymbeline:* Shakespeare's Valediction to Rome." *Roman Images.* Ed. Annabel Patterson. Selected Papers From the English Institute, 1982, n.s., n. 8. Baltimore: Johns Hopkins UP, 1984.

Modleski, Tania. *The Women Who Knew Too Much.* New York: Methuen, 1988.

Morgan, Irvonwry. *The Godly Preachers of the Elizabethan Church.* London: Epworth P, 1965.

Morgan, John. *Godly Learning.* Cambridge: Cambridge UP, 1986.

Mullaney, Steven. *The Place of the Stage.* Chicago: Chicago UP, 1988.

————. "Mourning and Misogyny: Hamlet, The Revenger's Tragedy, and the Final Progress of Elizabeth I, 1600–1607." *Shakespeare Quarterly* 45 (1994): 139–62.

Mullany, Peter. "Religion in Massinger and Dekker's *The Virgin Martyr.*" *Komos* 2 (1970): 89–97.

Murray, Peter B. *A Study of Cyril Tourneur.* Philadelphia: U of Pennsylvania P, 1964.

Neale, J. F. *Essays in Elizabethan History.* London: Cape, 1958.

Neely, Carol Thomas. *Broken Nuptials in Shakespeare's Plays.* New Haven: Yale UP, 1985.

Newman, Barbara. *Sister of Wisdom.* Berkeley: U of California P, 1987.

Newman, Jane O. "'And Let Mild Women to Him Lose Their Mildness': Philomela, Female Violence, and Shakespeare's *The Rape of Lucrece.*" *Shakespeare Quarterly* 45 (1994): 304–26.

Newman, Karen. *Fashioning Femininity and English Renaissance Drama.* Women in Culture and Society. Chicago: U of Chicago P, 1991.

Norgaard, Holger. "Never Wrong But With Just Cause." *English Studies* 45 (1964): 137–41.

Nutall, A. D. "*Measure for Measure:* Quid pro Quo?" *Shakespeare Studies* 4 (1968): 231–51.

————. "*Measure for Measure:* The Bed-trick." *Shakespeare Survey* 28 (1975): 51–56.

Ogilivie, R. M. *A Commentary on Livy Books 1–5.* Oxford: Clarendon P, 1965.

Oliphant, E. H. C. *Shakespeare and His Fellow Dramatists.* 2 vols. New York: Prentice Hall, 1929.

Orlin, Lena Cowen. *Private Matters and Public Culture in Post-Reformation England.* Ithaca: Cornell UP, 1994.

Otten, Elizabeth Spalding. "Massinger's Sexual Society." Levin and Robertson 137–56.

Ovid. *The Metamorphoses.* Trans. Arthur Golding. *Shakespeare's Ovid.* Ed. W. H. D. Rouse. London: Centaur, 1961.

Oxford English Dictionary. 1st ed. 1933.

Ozment, Stephen. *When Fathers Ruled.* Cambridge, MA: Harvard UP, 1983.

Pagels, Elaine. *Adam and Eve and the Serpent.* 1988. New York: Vintage-Random House, 1989.

Painter, William. *The Palace of Pleasure.* 1890. 3 vols. New York: Dover, 1966.

Paster, Gail Kern. *The Body Embarrassed.* Ithaca: Cornell UP, 1993.

Pearse, Nancy Cotton. *John Fletcher's Chastity Plays.* Lewisberg: Bucknell UP, 1973.

Peterson, Joseph Martin. *The Dorothea Legend.* Heidelberg: Buch-und Kunstdruck-eriei Rossler, 1910.

Petrarca, Francesco. *Opere.* Florence: Sansoni, 1975.

Pettie, George. *A Petite Pallace Of Pettie His Pleasure.* 1576. Ed. I. Gollancz. 2 vols. London: Chatto, 1908.

Plautus. *The Rope.* [*Rudens.*] *The Rope and Other Plays.* Trans. E. F. Watling. Harmondsworth: Penguin, 1984. 87–156.

Pollitt, Katha. "Violence in a Man's World." *Reasonable Creatures.* New York: Knopf, 1994. 26–30.

Porter, Roy. "Rape—Does it have a Historical Meaning?" Tomaselli and Porter 216–36.

Porterfield, Amanda. *Female Piety in Puritan New England.* New York: OUP, 1992.

Post, J. B. "Ravishment of Women and the Statutes of Westminster." *Legal Records and the Historian.* Ed. J. H. Baker. London: Royal Historical Society, 1978. 150–64.

———. "Sir Thomas West and the Statue of Rapes, 1382." *Rape and the Criminal Justice System.* Ed. J. Temkin. Aldershot: Dartmouth, 1995. 177–183.

Potter, Mary. "Gender Equality and Gender Hierarchy in Calvin's Theology." *Signs* 11 (1986): 725–39.

Powell, Nicholas. *Fuseli: The Nightmare.* Art in Context. New York: Viking, 1972.

Price, George R. *Thomas Dekker.* New York: Twayne, 1969.

Prior, Mary, ed. *Women in English Society 1500–1800.* London: Methuen, 1985.

Procopius of Caesarea. *Procopius.* Trans. H. B. Dewing. 7 vols. Loeb Classical Library. London: Heinemann, 1914–1940.

Prudentius. *Prudentius.* Trans. H. J. Thomson. 2 vols. Loeb Classical Library. London: Heinemann, 1949.

Randall, Dale B. J. "Some Observations on the Theme of Chastity in *The Changeling.*" *English Literary Renaissance* 14 (1984): 347–66.

Reuther, Rosemary Radford. "Misogynism and Virginal Feminism in the Fathers of the Church." *Religion and Sexism.* Ed. R. R. Reuther. New York: Simon and Schuster, 1974. 150–83.

Rhoads, Rebecca G. Introduction. *Two Noble Ladies.* i-ix.

Richardson, Samuel. *Clarissa.* 1747–48. Ed. Angus Ross. Harmondsworth: Penguin, 1985.

Richlin, Amy. "Reading Ovid's Rapes." Richlin, *Pornography and Representation in Greece and Rome* 158–79.

———, ed. *Pornography and Representation in Greece and Rome.* Oxford: Oxford UP, 1992.

Riefer, Marcia. "'Instruments of Some Mightier Member': The Constriction of Female Power in *Measure for Measure.*" *Shakespeare Quarterly* 35 (1984): 157–69.

Roberts, Anna, ed. *Violence Against Women in Medieval Texts.* Gainesville: UP of Florida, 1998.

Roberts, Sasha. "Editing Sexuality, Narrative and Authorship: The Altered Texts of Shakespeare's *Lucrece.*" *Texts and Cultural Change in Early Modern England.* Ed. Cedric C. Brown and Arthur F. Marotti. Early Modern Literature in History. New York: St. Martin's, 1997. 124–152.

Robertson, Karen. "Chastity and Justice in *The Revenger's Tragedy.*" Levin and Robertson 215–36.

Rogers, Katharine M. *The Troublesome Helpmate.* Seattle: U of Washington P, 1966.

Roper, Lyndal. *Oedipus and the Devil.* London: Routledge, 1994.

Rose, H. J. *A Handbook of Greek Mythology.* 1928. University Paperback. London: Methuen, 1964.

Rouselle, Aline. *Porneia.* Trans. Felicia Pheasant. Oxford: Blackwell, 1988.

Rowley, William. *William Rowley His All's Lost By Lust, and A Shoemaker, a Gentleman.* Ed. Charles Stork. University of Pennsylvania Series in Philology and Literature 13. Philadelphia: U of Pennsylvania P, 1910.

———and Thomas Middleton. *The Spanish Gipsie, and All's Lost By Lust, by Thomas Middleton and William Rowley.* Ed. Edgar C. Morris. Belles Lettres Series. Boston: Heath, 1908.

Rubin, Gayle. "The Traffic in Women: Notes on the 'Political Economy' of Sex." *Toward an Anthropology of Women.* Ed. Rayna R. Reiter. New York: Monthly Review, 1975. 157–210.

Sanday, Peggy Reeves. *Fraternity Gang Rape.* New York: New York UP, 1990.

———. "Rape and the Silencing of the Feminine." Tomaselli and Porter 84–101.

Schanzer, Ernst. "The Marriage-Contracts in *Measure for Measure.*" *Shakespeare Survey* 13 (1960): 81–9.

Schochet, Gordon. "Patriarchalism, Politics and Mass Attitudes in Stuart England." *Historical Journal* 12 (1969): 413–41.

Schoenbaum, Samuel. *Middleton's Tragedies.* New York: Columbia UP, 1955.

Schulenburg, Jane Tibbetts. "The Heroics of Virginity: Brides of Christ and Sacrificial Mutilation." *Women in the Middle Ages and the Renaissance.* Ed. Mary Beth Rose. Syracuse: Syracuse UP, 1986. 29–72.

Scully, Diana. *Understanding Sexual Violence.* Perspectives on Gender 3. Boston: Unwin, 1990.

———and Joseph Marolla. "Rape and Vocabularies of Motive." Burgess 1: 294–312.

The Second Maiden's Tragedy. Ed. Anne Lancashire. Revels Plays. Manchester: Manchester UP, 1978.

Sedgwick, Eve Kosofsky. *English Literature and Male Homosocial Desire.* Gender and Culture. New York: Columbia UP, 1985.

Semper, I. J. "The Jacobean Theatre Through the Eyes of Catholic Clerics." *Shakespeare Quarterly* 3 (1952): 46–51.

Sexton, Joyce H. *The Slandered Woman in Shakespeare.* English Literary Studies. Victoria: U of Victoria, 1978.

226   *Sexual Violence on the Jacobean Stage*

Shakespeare, William. *Cymbeline.* Ed. C. M. Ingleby. London, 1886.

———. *Cymbeline.* Ed. J. M. Nosworthy. Arden Shakespeare. 1955. University Paperback. London: Methuen, 1980.

———. *Cymbeline.* Literary Consultant: John Wilders. B.B.C. TV Shakespeare. London: B.B.C., 1983.

———. *Henry IV, Part I.* Ed. P. H. Davison. New Penguin Shakespeare. London: Penguin, 1968.

———. *Lucrece.* Ed. Sidney Lee. Oxford: Clarendon P, 1905.

———. *Measure For Measure.* Ed. Mark Eccles. New Variorum. New York: M.L.A., 1980.

———. *Measure For Measure.* Ed. J. W. Lever. Arden Shakespeare. 1965. University Paperback. London: Methuen, 1967.

———. *Pericles.* Ed. Philip Edwards. New Penguin Shakespeare. Harmondsworth: Penguin, 1976.

———. *Pericles.* Ed. F. D. Hoeniger. Arden Shakespeare. London: Methuen, 1963.

———. *The Poems.* Ed. F. T. Prince. Arden Shakespeare. 1960. University Paperback. London: Methuen, 1969.

———. *The Tempest.* Ed. Frank Kermode. Arden Shakespeare. 1954. University Paperback. London: Methuen, 1964.

———. *Titus Andronicus.* Ed. Eugene M. Waith. Oxford Shakespeare. Oxford: Clarendon P, 1984.

———. *Titus Andronicus.* Literary Consultant, John Wilders. B.B.C. TV Shakespeare. London: B.B.C., 1986.

Sharpe, J. A. *Crime in Early Modern England 1550–1750.* London: Longman, 1984.

———. *Defamation and Sexual Slander in Early Modern England.* Borthwick Papers No. 58. York: U of York, [1980].

———. "Witchcraft and Women in Seventeenth-Century England: Some Northern Evidence." *Continuity and Change* 6 (1991): 179–200.

Shirley, Henry. *The Martyred Soldier.* London, 1638.

———. *The Martyred Soldier. A Collection of Old English Plays.* Ed. A. H. Bullen. 4 vols. London, 1882–89. 1: 165–254.

Shorter, Edward. "On Writing the History of Rape." *Signs* 3 (1977): 471–82.

Sidney, Sir Philip. *The Countess of Pembroke's Arcadia.* Ed. Maurice Evans. Harmondsworth: Penguin, 1977.

Siemon, James E. "Noble Virtue in *Cymbeline.*" *Shakespeare Survey* 29 (1976): 51–61.

Sisson, C. J. "Shakespeare's Quartos as Prompt Copies." *Review of English Studies* 18 (1942): 135–43.

Slater, Miriam. *Family Life in the Seventeenth Century.* London: Routledge, 1984.

Smith, Bruce. *Homosexual Desire in Shakespeare's England.* 1991. Paperback ed. Chicago: U of Chicago P, 1994.

Smith, Hilda. "Gynecology and Ideology in Seventeenth-Century England." *Liberating Women's History.* Ed. Berenice A. Carroll. Urbana: U of Illinois P, 1976. 97–114.

———. *Reason's Disciples.* Urbana: U of Illinois P, 1982.

Smith, John Hazel. "Charles the Bold and the German Background of the 'Monstrous Ransom' Story." *Philological Quarterly* 51 (1972): 380–93.

Sommerville, J. P. *Politics and Ideology in England 1603–1640.* London: Longman, 1986.

Spenser, Edmund. *The Faerie Queene.* Ed. A. C. Hamilton. London: Longman, 1977.

Stallybrass, Peter. "Patriarchal Territories: The Body Enclosed." *Rewriting the Renaissance.* Ed. Margaret W. Ferguson, Maureen Quilligan and Nancy J. Vickers. Chicago: U of Chicago P, 1986. 123–42.

Steppat, Michael Payne. "John Webster's *Appius and Virginia.*" *American Notes and Queries* 20 (1982): 101.

Stimpson, Catherine R. Foreword. *Fashioning Femininity and English Renaissance Drama.* By Karen Newman. xi–xiv.

———. "Shakespeare and the Soil of Rape." *The Woman's Part.* Ed. Carolyn Ruth Swift, Gayle Greene and Carol Thomas Neely. Urbana: U of Illinois P, 1980. 56–64.

Stockton, Sharon. "The 'Broken Rib of Mankind': The Sociopolitical Function of the Scapegoat in *The Changeling.*" *Papers on Language and Literature* 26 (1990): 459–77.

Stone, Lawrence. *The Causes of the English Revolution 1529–1640.* 1942. Ark edition. London: Routledge-Methuen, 1986.

———. *The Crisis of the Aristocracy 1558–1641.* Oxford: Clarendon, 1965.

———. *The Family, Sex and Marriage in England, 1500–1800.* London: Weidenfield, 1977.

Strong, Roy. *The Cult of Elizabeth.* London: Thames, 1977.

Styan, J. L. *Shakespeare's Stagecraft.* Cambridge: Cambridge UP, 1967.

Suleiman, Susan Rubin, ed. *The Female Body in Western Culture.* Cambridge, MA: Harvard UP, 1985.

Swander, Homer D. "*Cymbeline* and the 'Blameless Hero.'" *English Literary History* (1964): 259–70.

Tacitus. *Annals.* Trans. John Jackson. *Tacitus.* Vols. 3–5. Loeb Classical Library. London: Heinemann, 1914.

Tatius, Achilles. *The History of Clitophon and Leucippe.* 1597. Trans. W. B. Amsterdam: Theatrum, 1977.

Tawney, R. H. *Business and Politics Under James 1.* Cambridge: Cambridge UP, 1958.

Taylor, Michael. "The Pastoral Reckoning in *Cymbeline.*" *Shakespeare Survey* 36 (1983): 97–106.

Tertullian. "The Apparel of Women." Trans. Edwin A. Quain. *Disciplinary, Moral and Ascetical Works.* The Fathers of the Church 40. New York: Fathers of the Church, 1959.

———. *Opera. Patrilogiae Latinae* 1–2. Paris, 1879.

Thirsk, Joan. Foreword. *Women in English Society 1500–1800.* Ed. Mary Prior. 1–21.

Thomas, Keith. "The Double Standard." *Journal of the History of Ideas* 20 (1959): 195–216.

———. "The Puritans and Adultery: the Act of 1650 Reconsidered." *Puritans and Revolutionaries.* Ed. Donald Pennington and Keith Thomas. Oxford: Clarendon P, 1978. 257–82.

———. *Religion and the Decline of Magic.* 1971. Harmondsworth: Peregrine-Penguin, 1978.

————. "Women in the Civil War Sects." *Past and Present* 13 (1958): 42–62.

Thomas, Vivian. *The Moral Universe of Shakespeare's Problem Plays.* Totawa: Barnes and Noble, 1987.

Thompson, Ann. "Philomel in *Titus Andronicus* and *Cymbeline.*" *Shakespeare Survey* 31 (1978): 23–32.

Thornhill, Randy, Nancy Thornhill, and Gerard A. Dizinno. "The Biology of Rape." Tomaselli and Porter 102–121.

Tillyard, E. M. W. "*Measure for Measure.*" Geckle 98–102.

Tilney, Edmund. *The Flower of Friendshippe.* London, 1568.

Todd, Barbara J. "The Re-Marrying Widow: A Stereotype Reconsidered." Prior, 54–92.

Todd, Margo. *Christian Humanism and the Puritan Social Order.* Cambridge: CUP, 1987.

Tomaselli, Sylvana and Roy Porter. *Rape.* Oxford: Blackwell, 1986.

Toner, Barbara. *The Facts of Rape.* London: Hutchinson, 1977.

Tong, Rosemarie. *Women, Sex, and the Law.* Totawa: Rowman, 1984.

Tourneur, Cyril. *The Atheist's Tragedy.* Ed. Irving Ribner. Revels Plays. London: Methuen, 1964.

————. *The Revenger's Tragedy.* Ed. Brian Gibbons. New Mermaids. London: Benn, 1967.

The Two Noble Ladies. Ed. Rebecca G. Rhoads. Malone Society Reprints 61. Oxford: Oxford UP, 1930.

Underdown, David. *Fire From Heaven.* 1992. London: HarperCollins-Fontana, 1993.

————. "The Taming of the Scold: the Enforcement of Patriarchal Authority in Early Modern England." Fletcher and Stevenson 116–36.

d'Urfé, Honoré. *L'Astrée.* 5 vols. Geneva: Slatkine, 1966.

Vega, Lope de. *Fuente Ovejuna.* Trans. William E. Colford. Woodbury, NY: Barron's, 1969.

Vermeersch, A. "Virginity." *The Catholic Encyclopedia.* Ed. C. G. Heibermann et al. 15 vols. New York: Encyclopedia P, 1913.

Vickers, Nancy J. "'This Heraldry in Lucrece' Face'." Suleiman 209–22.

Victor of Vita, St. *The Memorable and Tragicall History of the Persecution in Africke.* [Trans. Ralph Buckland.] N.p., 1605.

Vogelman, Lloyd. *The Sexual Face of Violence.* Johannesburg: Ravan P, 1990.

Walker, Garthine. "Rereading Rape and Sexual Violence in Early Modern England." *Gender and History* 10 (1998): 1–25.

Wall, Alison. "Elizabethan Precept and Feminine Practice: the Thynne Family of Longleat." *History* 75 (1990): 23–38.

Walzer, Michael. *The Revolution of the Saints.* 1965. New York: Athenaeum, 1973.

Warner, Marina. *Alone of All Her Sex.* 1976. New York: Vintage-Random House, 1983.

————. "The Wronged Daughter." *Grand Street* 7.3 (1988): 143–63.

A Warning For Fair Women. Ed. Charles Dale Cannon. Studies in English Literature 86. The Hague: Mouton, 1975.

Warren, Roger. *Cymbeline*. Shakespeare in Performance. Manchester: Manchester UP, 1989.

——. "Theatrical Virtuosity and Poetic Complexity in *Cymbeline*." *Shakespeare Survey* 29 (1976): 41–49.

The Wars of Cyrus. Ed. James P. Brawner. *Illinois Studies in Language and Literature* 28.3–4 (1942).

Watt, David. *The New Offences Against the Person*. Toronto: Butterworths, [1984].

Watt, Tessa. *Cheap Print and Popular Piety 1550–1640*. Cambridge: Cambrige UP, 1991.

Webster, John. *The Complete Works of John Webster*. Ed. F. L. Lucas. 3 vols. 1927. New York: Gordian, 1966.

——. *The Duchess of Malfi*. Ed. John Russell Brown. Revels Plays. 1964. Manchester: Manchester UP, 1976.

Welles, Marcia L. "Violence Disguised: Representation of Rape in Cervantes' 'La Fuerza de la Sangre.'" *Journal of Hispanic Philology* 13 (1989): 240–52.

Wentersdorf, Karl P. "The Marriage Contracts in *Measure for Measure:* A Consideration." *Shakespeare Survey* 32 (1979): 129–44.

Whately, William. *A Bride-Bush, or a Wedding Sermon*. 1617. Amsterdam: Theatrum, 1975.

White, Helen C. *Tudor Books of Saints and Martyrs*. Madison: U of Wisconsin P, 1963.

White, Paul Whitfield. "Theater and Religious Culture." *A New History of Early English Drama*. Ed. David Scott Kastan and John D. Cox. New York: Columbia UP, 1997. 133–52.

Wickham, Glynne. *Early English Stages 1300 to 1660*. 3 vols. New York: Columbia UP, 1959–81.

Wiesner, Merry E. "Spinning Out Capital: Women's Work in the Early Modern Economy." *Becoming Visible*. Ed. Renate Bridenthal, Claudia Koontz and Susan Stuard. 2nd ed. Boston: Mifflin, 1987. 222–49.

Williams, Carolyn. "Shakespeare and the Meanings of Rape." *Yearbook of English Studies* 23 (1993): 93–110.

Willis, Deborah. "The Monarch and the Sacred: Shakespeare and the Ceremony for the Healing of the King's Evil." Woodbridge and Berry 147–168.

Wilson, John. *The English Martyrologie*. 1608. Amsterdam: Da Capo-Theatrum, 1970.

Wilson, Stephen, ed. *Saints and Their Cults*. Cambridge: Cambridge UP, 1983.

Winston, Matthew. "'Craft Against Vice': Morality Play Elements in *Measure for Measure.*" *Shakespeare Studies* 14 (1981): 229–48.

Wolfthal, Diane. *Images of Rape*. Cambridge: Cambridge UP, 1999.

Woodbridge, Linda. *The Scythe of Saturn*. Urbana: U of Illinois P, 1994.

——. *Women and the English Renaissance*. 1984. Illini Books. Urbana: U of Illinois P, 1986.

——and Edward Berry, ed. *True Rites and Maimed Rites*. Illini Books. Urbana: U of Illinois P, 1992.

Wright, C. E. "The Dispersal of the Libraries in the Sixteenth Century." Francis Wormald and C. E. Wright, ed. *The English Library Before 1700.* London: U of London-Athlone P, 1958. 148–75.

Wright, Louis B. *Middle Class Culture in Elizabethan England.* New York: Cornell UP, 1958.

Wrightson, Keith. *English Society 1850–1680.* London: Unwin, 1982.

Wynne-Davies, Marion. "'The Swallowing Womb'": Consumed and Consuming Women in *Titus Andronicus.*" *The Matter of Difference.* Ed. Valerie Wayne. Ithaca, NY: Cornell UP, 1991. 129–151.

Xenophon. *An Ephesian Tale. Three Greek Romances.* Trans. Moses Hadas. Indianapolis: Bobbs-Merrill, 1964. 69–126.

Xenophon. *The School of Cyrus. [Cyropaedia.]* 1567. Trans. William Barker. Ed. James Tatum. The Renaissance Imagination 37. New York: Garland, 1987.

Yates, Frances A. *Astraea.* 1975. Ark Paperbacks. London: Ark-Routledge, 1985.

Index

Adelman, J., 201n12, 202n25
All's Lost By Lust (Rowley), 23, 82,
    106–11, 123, 158, 159–60,
    168n35, 182n41
Apius and Virginia (R. B.), 63–4
Appius and Virginia (Webster), 23,
    74–81, 104, 159, 196n65,
    206n44, 206n46
assault, sexual, 2, 7
    as chastity test, 7, 10
    iconography of, 48–9, 50
    incestuous, 34, 48–9, 57–8
    necrophilous, 77–9
    victim of,
        converts assailants/suitors, 38–9,
            42, 56–7, 59
        erotic torture of, 51–2
        sanctification of, 26–31, 33–59,
            75–6
        pre-emptive murder of, 62, 63–4,
            75–9
        pre-emptive suicide of, 29, 45
    see also rape; saints
Atheist's Tragedy, The (Tourneur), 7,
    165–6n21, 189n10, 208–9n2
Atwood, M., 154

Bach, R. A., 8, 124
Baines, B., 123, 153, 163n1, 164n6,
    166n22, 186–7n20
Bashar, N., 5, 164n6, 165n18
Basil of Ancyra, 30, 59
Beaumont, F., 8

Bennett, J., 168n38
Benson, P. J., 171n59, 171n61
Bentley, T., 18
Bergeron, D. M., 190n13
Berry, P., 187n22
Blanchard, W. H., 206n42
Boccacio, G., 86, 87
Boke of Justices of Peas, The, 2, 7
Bonduca (Fletcher), 23, 82, 116–22,
    123, 158, 159
Book of Common Prayer, The, 13, 40, 88
Boose, L. E., 202n24
Bowers, F. T., 195n51
Bracton, H., 3
Bradbrook, M. C., 39
Breitenberg, M., 7, 168n37
Briggs, J., 198n81
Briggs, R., 21
Brittin, N., 207n49
Brown, L., 182n42
Brown, P., 26
Brownlow, F. W., 177n11, 178n19
Brownmiller, S., 4, 7, 190n13
Bruster, D., 172n68
Buckland, R., 41
Bullough, G., 37
Burelbach, F. M., Jr., 207n50
Burks, D., 2, 3, 151–2, 163n1, 163n5
Burt, J. R., 196n61, 198n75
Bussy D'Ambois (Chapman), 48,
    208n58

Cahn, S., 169n42

Calder-Marshall, A., 186n12
Calvin, J., 14
Camelot, P. T., 29
Carter, A., 33
Carter, J. M., 165n19
Cary, E., Lady Falkland, 192–3n31
Cervantes Saavedra, M. de, 143–4,
    150, 207n54, 207–8n57,
    208n60
Chamberlain, J., 1, 176n25
Changeling, The (Middleton/Rowley),
    32, 151–2, 168n36, 197n72
Chapman, G., 48, 208n58
charivaris, 24, 33
chastity, 1–2, 10, 17
    anxiety about, 17–19
    and male property, 79, 131–32, 157
    eroticization of, 7
    exemplars of, 20
    material nature of, 88, 115, 155,
        157
    see also rape; saints; sexual assault
Chaucer, G., 63, 186n17
Chaytor, M., 6, 164n6, 165n15,
    168n38
Children of the Chapel, 182n47
Cinthio, G., 125–7, 153–4, 208n59
Clark, Alice, 12
Clark, L., and D. Lewis, 155, 164n9
Clark, Sandra, 120, 199n94
Clark, Stuart, 21
Claudian, 205n39
Cockburn, J. S., 165n14
Coleridge, S. T., 1
Couliano, I. P., 16
Crane, M. T., 184n54
Crawford, P., 169n42
Cressy, D., 17, 171n64
cursing, 16, 67, 107–9
Cymbeline (Shakespeare), 23, 55–6,
    81–2, 85–93, 97, 98, 156,
    157, 187n28

Dante, 206n43
Davies, K. M., 14

Davis, N. Z., 16
Dawson, A. B., 163n1
defamation, 18–19, 209n4
Dekker, T., 23, 33–4, 46–53, 58–9,
    157, 207n49
Detmer, E., 2, 163n1
De Voragine, J., 26, 27, 30, 37, 44, 55
Dick of Devonshire (anon.), 200n3
Diehl, H., 153
Digangi, M., 163n2
Dinshaw, C., 167n29
Dolan, F. E., 16, 18, 109, 172n67
Donaldson, I., 73
Douglas, M., 9
Dubinsky, K., 164n12
Dubrow, H., 170n57, 189n10
Duchess of Malfi, The (Webster), 135

Easton, M., 174n8
Edwards, P., 40, 176–7n1, 202n18
Eisenstein, E., 169n47
Empson, W., 186n18, 186n20,
    202n23
Erasmus, D., 25
Erickson, A. L., 3, 13
Eusebius, 173–4n5, 175n15, 192n30
Evans, B., 203n25
Ewbank [Ekeblad], I., 40
Ezell, M. J. M., 169n44

Faithful Shepherdess, The (Fletcher),
    191n18
Fenton, G., 61
Field, N., 23, 123–5,132–43, 153–4,
    155, 156, 158, 160,
    167–8n35, 183n50
Findlay, A., 163n2
Finke, L. A., 198n77
Fletcher, J., 8, 23, 82, 100–6 116–22,
    123–5, 132–43, 153–4, 155,
    156, 157–8, 159, 160,
    166n27, 167–8n35, 168n36,
    191n18, 196n64, 197n70,
    201n8, 206n46
Ford, J., 181–2n39, 207n49, 208n58

Foxe, J., 31, 175–6nn21–22, 178–9n19, 182n46, 192n30, 193n38
Frye, N., 202n25
Furies (*Erinyes*), 110, 135, 136, 138, 158

Garber, M., 202n25
Gasper, J., 182n43
gender, anxiety about, 12–22
Gerard, J., 176n23
Gesner, C., 189n9
Ghost of Lucrece, The (Middleton), 187n21
Girard, R., 7–8
Gless, D., 201–2n14
Goffe, T., 183n48
Golden Legend, The, see de Voragine, J.
Goolden, P., 35
Gorfain, P., 202n16
Gossett, S., 5, 10–11, 79, 123–4, 152, 154, 163n1, 195n51
Gower, J., 34, 35, 36, 38
Gowing, L., 18, 19, 209n4
Gramsci, A., 173n70
Granville-Barker, H., 86, 90
Green, P. D., 199nn93–4, 200n96
Greene, R., 192n28
Griffin, S., 4
Gurr, A., 172n69, 177n5

Haber, J., 158
hagiography, 25–32
    influence of on drama, 32, 37–59, 63–4, 75, 83–4, 91–2, 93–99, 100–1, 107, 114–130
    see also saints, *and under individual names*
Hale, Sir M., 2, 3, 5, 6, 199n88
Hali Meidenhad (anon.), 31
Halkett, J., 14, 15
Haller, W. and M., 14
Harris, V., 20, 21
Harris, W. V., 188n34
Heal, F., 21

Helms, L., 188n30
Hengist, King of Kent (Middleton), 23, 82, 112–16, 122, 155, 156, 157, 168n36, 201n13, 204n33
Hester, M., 21, 170n53
Heywood, T., 23, 55–6, 68–74, 156, 158, 159, 160, 166–7n28, 167–8n35, 184–5n3, 188n32
Hildegard of Bingen, 174n10
Hill, C., 14
Historia Apollonii Regis Tyrii, 34, 35, 37, 38, 183n49
Holinshed, R., 112, 116, 117, 198n80, 198n82, 199n91
Houlbrooke, R., 14
Howard, J., 172n69
Hrotsvitha, 29, 193n35
Hutson, L., 168n37

Ingleby, C. M., 190n14
Ingram, M., 16, 19, 170nn50–1

James I and VI, 15, 16, 21, 33
Jardine, L., 132, 168n37, 191n22, 202n25
Jensen, P., 176n24
Jonson, B., 7, 8, 21, 23, 32, 33, 54, 205n37, 208–9n2
Joplin, P. K., 9, 167nn29–30, 167nn32–33, 181n38

Kahn, C., 66, 163n1, 186n13, 187n22
King's Men, 86, 93, 156, 188n1, 194n43, 194n45, 198n79, 199n90, 205n36
Kirsch, A., 189n9, 191n20
Kistner, A. L. and M. K., 145, 146, 193n38, 207n52
Knight, G. W., 202–3n25
Kortekaas, G. A. A., 37
Kyd, T., 208n58

Lady Elizabeth's Men, 198n79, 207n49
Lamb, M. E., 9, 108, 186n14

Lancashire, A., 93, 94, 97, 99,
    194nn43–4, 198n83
Larner, C., 170n52, 174n6
Lascelles, M., 125
Lawes Resolutions of Womens Rights, The
    (T. E.), 13
Levin, R., 22, 193n37
Lévi-Provencal, É., 106
Lindsay, R., 190–1n17
Lives of Women Saints, 28
Livy, 61–2, 69, 73–4, 185–6n10
Lovelace, R., 195n53
Lucas, F. L., 74
Lucrece (Shakespeare), 55–6, 65–8, 70,
    158, 167n31, 167n33,
    184n54, 189n10, 190n13,
    194n47, 195n49, 207n55,
    209n6
Lucretia/Lucrece (legendary), 9, 61–3,
    82–3, 167n30
Luther, M., 13–14

Mac-, *see also* Mc-
MacFarlane, A., 14, 170n53
Machiavelli, N., 81
Mackinnon, C. A., 2
Maclean, I., 12
Marcus, L. S., 5
marriage, Protestant, 14–15
Marolla, J., 6, 164n9
Marsden, J. I., 163n1, 165n20, 181n39
Martyred Soldier, The (Shirley), 23,
    33–4, 41–6, 49–50, 56, 58–9,
    71, 157
martyrs, virgin *see* saints, *and under
    individual names*
Mary, Blessed Virgin, 26, 30–1, 39–40
Mask, A (Milton), 7, 183n51, 205n40
Massey, D., 132, 204n30
Massinger, Philip, 10, 23, 33–4, 46–53,
    58–9, 123–5, 132–43, 153–4,
    155, 156, 157,158, 160
McLuskie, K., 11, 22, 130
Measure For Measure (Shakespeare), 2,
    23, 125–32, 153–4, 156, 160

Medusa, 9–10, 65, 80, 138, 158
Middleton, T., 7, 23, 32, 112–16,
    122–4, 143–54, 155, 156,
    157, 158, 159, 165n21,
    1678nn35–6, 184n54, 184n1,
    187n21, 197n72, 200n3,
    201n13, 204n33
    *see also Revenger's Tragedy, The;
        Second Maiden's Tragedy, The*
Mikalachki, J., 189n8, 199nn92–3,
    200n97
Miles, M. R., 170n54
Milton, J., 7, 183n51, 205n40
Miola, R. S., 190n13
Moshinsky, E., 190n17
Mullaney, S., 178n13, 189n5

Neely, C. T., 88, 177n9
Newman, J. O., 9, 163n1, 167n31
Newman, K., 14
Nosworthy, J. M., 89
Nutall, A. D., 201n13, 202n18

Ogilivie, R. M., 61, 185n5
Orlin, L. C., 14, 19, 171–2n66,
    196n59
Othello (Shakespeare), 1, 55–6,
    172n66, 196n59
Otten, E. S., 10
Ovid, 9–10, 54, 65, 138, 186n17,
    196n66
Ozment, S., 169n44

parricide, 48, 62, 63–4, 69–70, 76–9
Paster, G. K., 17
Pearse, N. C., 17, 185n6, 207n48
Pericles (Shakespeare), 23, 33–41, 55–6,
    58–9, 156, 157, 183n49
Peterson, J. M., 49
Petrarch, F., 39–40
Philaster (Beaumont/Fletcher), 8,
    166n27
Philomela, 9, 80, 101, 108, 111, 158,
    191–2n19, 196n66
Porter, R., 4, 5, 164n7

Porterfield, A., 169n41
Prince, F. T., 187nn21–2
Prince Charles's Men, 195n56, 198n79
Procopius of Caesarea, 100, 103, 104, 105
Promos and Cassandra (Whetstone), 125, 127–8, 168n36, 202n19
Prudentius, 26

Queen Anne's Men, 187n23
Queen of Corinth, The (Fletcher/Massinger/Field), 23, 123–4, 132–43, 153–4, 155, 156, 158, 159, 167–8n35, 168n36, 201n8

rape
and male homosocial bonds, 7, 8, 49–50, 73–4, 77–80, 81, 90, 124, 133–4, 136, 201n8
and patriarchy, 4–9
as source of bawdy comedy, 72–3, 114–5
as theft, 2–3, 67, 79, 88
community's tolerance of, 5–6, 73, 124
exploited politically, 68, 73–74, 85
feminist theories of, 3–4
gang, 27, 50, 53–4, 136
incidence of, in early modern England, 5–6
legal aspects, 2–3, 5–6,155, 199n88
marital, 115, 155
myths of, 6–7, 26–30
prosecutions for, 5–6, 124
specular, 55–6, 66, 87–9
threat of, represented erotically, 65–6, 87–8
victim of
anger of, 9, 67, 126–7, 137, 159
assumption of blame by, 29, 67–8, 70, 137, 146, 157
blamed, 7, 35–6, 67, 81, 88, 157
demonization of, 65, 67–8, 70, 80, 82, 107–9, 111, 136–7,

117–22, 126–27, 156, 158, 159
loss of identity of, 10, 70, 136–7
marriage to rapist, 10–11, 123–28, 158–9
murdered, 65, 111
polluted/polluting, 9, 67, 137, 158, 159
sacrificial, 9, 61–4, 70, 71, 75–6
scapegoating of, 22, 67
suicidal, 34, 67–8, 82–3, 127, 145
typology of, 7
Rape of Lucrece, The (Heywood), 23, 55–6, 68–74, 77, 79, 80, 81
rapist
heroic, 74, 190n13
sympathetic, 143–6
typology of, 7
Revenger's Tragedy, The (attrib. Tourneur), 23, 81–5, 99, 112, 156, 159, 167–8n35, 192n29, 194nn39–40, 195n55, 204n32, 205n41
Rhoads, R. G., 53
Richardson, S., 123, 190n15, 201n9
Roberts, S., 185n6
Robertson, K., 82, 189n5
Roper, L., 209n5
Rowley, W., 23, 32, 82, 106–11, 123–4, 143–54, 151–2, 155, 158, 159, 160, 166n21, 168n35, 182n41
Rubin, G., 166n23

St. Agnes, 27, 37, 38, 49, 176n22
St. Aldhelm, 27, 173n5, 175n16
St. Ambrose, 25, 29, 175n16
St. Augustine, 13, 29, 174n14, 186n20, 203n26
St. Barbara, 26, 180n30, 181n36
St. Cyprian, 53
St. Dorothea, 46
St. Drusiana, 29, 45, 93,
St. Dympna, 57, 58
St. Eugenia, 26, 27, 178–9n19

St. Jerome, 175n16
St. Justina, 53, 88
St. Katherine, 44
St. Lucy, 27, 30
St. Sophronia, 93, 193n38
St. Ursula, 28
St. Victor of Vita, 41
St. Winifred, 27–8, 32
saints, 25–32
    cult of, in England, 31–2
    erotic torture of, 26–7
    exemplary chastity of, 26–31
    inviolability of, 28–30, 45
    subjected to brothel, 27, 29, 30,
        37–8
    symbolic rape of, 26–7
    see also hagiography; *and under*
        individual names
Sanday, P. R., 3–4, 164n9, 206n42
Schoenbaum, S., 192n29
Schulenburg, J. T., 175nn15–6
scolding, 16, 21
Scully, D., 6, 164n9
Second Maiden's Tragedy, The (anon.),
        22–3, 55, 57, 81, 93–100,
        112, 114, 115, 122, 134, 156,
        157, 159, 160, 177n4, 189n10
Sedgwick, E. K., 2
Semper, I. J., 179n25
Shakespeare, W., 8, 23, 33–41, 55–6,
        58–9, 64–8, 70, 76, 81–2,
        85–93, 97, 98, 156, 157, 158,
        167n31, 167n33, 183n49,
        184n54, 187n28, 199n87,
        194n47, 195n49, 197n71,
        197n74, 207n55, 209n6
Sharpe, J. A, 165n14
Shichtman, M., 198n77
Shirley, H., 23, 33–4, 41–6, 49–50, 56,
        58–9, 71, 157
Sidney, Sir P., 32, 189n11, 205n38,
        208n59
Siemon, J. E., 191nn19–20
Sophonisba (Marston), 32, 33, 48,
        189n10, 208–9n2

Spanish Gypsy, The (attrib.
        Middleton/Rowley), 23,
        123–4, 143–54, 155, 158–9,
        166n21, 167- 8n35
Spenser, E., 32, 86, 166n26, 189n10,
        191n18, 205n40
Stimpson, C., 12, 86, 163n1, 167n34,
        168n37
Stockton, S., 151
Stone, L., 14, 20–1, 169n42
Styan, J. L., 131

Tatius, A., 175n19
Taylor, M., 90–1
Tertullian, 26
Thomas, K., 16, 32, 108, 170n53
Thomas, V., 202n25
Thompson, A., 191n24
Tillyard, E. M. W., 203n27
Tilney, E., 18
'Tis Pity She's A Whore (Ford), 181n39,
        208n58
Titus Andronicus (Shakespeare), 8,
        64–5, 76, 85, 156, 157, 158,
        159, 199n87, 209n6
Todd, M., 169n45
Tourneur, C., 7, 165–6n21, 189n10,
        208–9n2
Twine, L., 34, 35, 37, 38
Two Noble Ladies, The (anon.), 23,
        33–4, 53–9, 157, 205n40

Underdown, D., 15, 16, 20, 171n63
Unnatural Combat, The (Massinger),
        182n40, 209n3
d'Urfé, H., 103, 105, 194n46, 195n51

Valentinian (Fletcher) 23, 82, 100–6,
        123, 157–8, 159, 168n35,
        196n64, 197n70, 206n46
Vickers, N. J., 163n1, 167n33, 186n19
Virgil, 205nn37–8
Virgin Martyr, The (Massinger/Dekker),
        23, 33–4, 46–53, 58–9, 157
Virginia (legendary), 61–3

virginity, 25–6, 29, 31
 defined by Catholic Church, 29
 in early Christian church, 25–6
Vives, J. L., 31
Volpone (Jonson), 2, 7, 8, 21, 23, 32,
 33, 54, 208–9n2

Walker, G., 6, 124
Walzer, M., 14, 15, 20
Warner, M., 26, 39, 181n36
Warning For Fair Women, A (anon.),
 18
Warren, R., 190n17, 191n20
Wars of Cyrus, The (anon.), 55–6
Watt, T., 170n55, 171n60
Webster, J., 23, 74–81, 104, 135, 159,
 196n65, 206n44, 206n46
Welles, M. L., 208n62
Whately, W., 14, 15, 16
Whetstone, G., 125, 127–8, 168n36,
 202n19

White, P. W., 32
Wickham, G., 181n34
Williams, C., 8, 163n1, 167n34,
 200n1
witchcraft, 16, 21, 108
Wolfthal, D., 181n38
women, 3, 7, 12–22, 33–34
 and capitalism, 12
 and Protestant theology, 13–15
 as male property, 3, 7
 as playgoers, 22
 disorderly, 15–17, 21, 22
 formal controversy about, 19–20
 legal disabilities of, 12–13
Women Beware Women (Middleton), 7,
 32, 184n54, 197n72, 200n3
Woodbridge, L., 19–20, 21, 22,
 168n37, 171n62, 171n66,
 184n54, 198n78, 209n6
Wright, L. B., 20, 169n48, 171n62
Wynne-Davies, M., 164nn6–7